DATE DUE

~~Feb 26 '73~~			
~~Jul 24 '74~~			
~~Dec 18 '74~~			
Mar 26 '77			
May 3 '77			
GAYLORD M-2			PRINTED IN U.S.A.

WEATHER AND LIFE

An Introduction to Biometeorology

WEATHER AND LIFE
An Introduction to Biometeorology

William P. Lowry
Oregon State University

ACADEMIC PRESS · NEW YORK and LONDON

ACADEMIC PRESS, INC.
111 Fifth Avenue, New York, New York 10003

United Kingdom Edition published by
ACADEMIC PRESS, INC. (LONDON) LTD.
Berkeley Square House, London, W.1.

LIBRARY OF CONGRESS CATALOG CARD NUMBER: 69-18334

PRINTED IN THE UNITED STATES OF AMERICA

CONTENTS

PREFACE

"THE DOMINANT TENDENCY in contemporary education is to teach man how to do things rather than how to exercise creative options. The dominant tendency of science is to emancipate man from doing things, enabling him to preside over more open time than he has known since ancient Greece. The combined result is that man has wondrous new options he is not prepared to recognize or enjoy."

Norman Cousins
Saturday Review
April 20, 1968

This book is based upon notes developed for a one-term course entitled Introduction to Biometeorology, and taught at Oregon State University since 1960. The students have been from a wide variety of backgrounds, and, as a group, have generally proved the unfortunate truth of what Mr. Cousins has to say.

My remark must not be taken as an indictment of students. Indeed they and most of their mentors are part of an "establishment" which has too long taught us all to see books and courses as collections of facts, formulas, references, recipes, and unassailable conclusions.

It is my hope that this book may be seen at least as a helping hand to those who are willing to develop the ability to communicate back and forth across the "boundaries" between biology and atmospheric science. At most it can be seen as a guide to thinking about parts of the way the world works—as a basis for exercising creative options, in Mr. Cousins' words. It may still be seen as a rigid collection of facts, but at least you should know of my hopes.

The first section treats the basic environmental factors of radiation, temperature and heat, moisture, and wind, more or less separately. The second section weaves these factors together in a fabric based on the energy budget, with a philosophical excursion into matters of data collection and instrumentation. The third section examines some of the more widely accepted biological concepts, using the energy budget as a point of reference. The fourth section treats the timely problems of urban climate and air pollution very briefly and descriptively.

Cross referencing in both text and problems is intended to help blend the four sections, but any combination of the four should be self-sustaining. To many persons already familiar with the energy-budget idea as used in a particular set of problems, the treatment here may seem overly broad and academic. Experience has taught, however, that students grasp the concept better when they know the common base from which the various special applications in the literature have come.

This book tries to make two points. First, the interactions between and among the various parts of an organism–environment system are extraordinarily complex. Second, despite the complexity there are ways of analyzing information so as to permit comprehension of the whole and to allow intelligent management of the system.

In the interest of developing a book which stops just after making these two points, but which includes "equal time" for both biologists and meteorologists (a thing no other book I am aware of does), I have purposely simplified and omitted arbitrarily. The emphasis is meant to be on concepts and the development of perspective. Thus, since the book does not aspire to compendium status, some topics are only mentioned. It does not have exhaustive bibliographies, and it may contain statements and judgments not wholly acceptable to other biometeorologists.

Whatever it is not, I intend that it will be direct, spare, internally consistent, and useful to a broad spectrum of interested readers.

Who knows? It may even be an inspiration to a handful, and they may help lead the way out of the dilemma posed by Norman Cousins.

W.P.L.

Corvallis, Oregon

Chapter 1

INTRODUCTION

The Nature of Biometeorology

BIOMETEOROLOGY HAS BEEN DEFINED by the International Society of Biometeorology as "the science which investigates the effects of geophysical and geochemical elements of the atmospheric environment on living organisms." For the purposes of this book, a definition indicating more clearly the effects operating in the opposite direction would be more appropriate: "the science which investigates the interrelationships of atmospheric and biological processes." Because there is so much more energy contained in the pure geophysical environment than in the mass of living material found in the physical matrix, we often think of the direction of controlling processes as being exclusively from the physical environment to the organism. While the results of biological processes may have relatively small effects on large scale atmospheric processes, on the microscale such effects are frequently very important, if not dominant. One has only to walk between rows of vigorously transpiring corn plants and experience the discomfort of the high humidity there to understand that the biological often determines the physical.

Terminology has come into use rather haphazardly in biometeorology, as it has in all other sciences during their formative years. There seems to be no way of telling very precisely the differences among terms such as "biometeorology," "bioclimatology," "physiological ecology," "environmental science," "environmental physiology," "aerobiology," and so on. To the uninitiated these will probably all sound pretty much alike, and even to the discerning specialist some will be equivalent to one or two others. There are differences, however, which can be pointed out as an aid in further defining the science.

Of the six terms listed, three have some reference to the atmosphere implicit in them: biometeorology, bioclimatology, and aerobiology. The other three do not. Aerobiology is commonly used to refer to the study of the composition and the ecology of airborne microorganismal populations. Biometeorology usually refers to interactions between atmospheric and biological processes involving only short periods of time, of the order of a few hours or a few days. Bioclimatology in its broad sense refers to such interactions on a longer time scale, and in a more restricted sense to an approach to research which will be examined presently.

Environmental science is probably best considered an all-encompassing term, including the full spectrum from plant uptake of trace elements in the soil to social psychology as related to space medicine. Physiological ecology is slightly more restrictive in that it seems to be concerned with the effects of the physical environment on individual biological organisms, therby laying aside for the most part interest in the effects of other kinds of organisms on individuals or in interactions of several organisms of the same kind.

This leaves for discussion the term environmental physiology and the more restricted meaning of bioclimatology. Environmental physiology has come to refer to something apparently quite similar to physiological ecology, but it seems to imply more specifically a particular experimental approach to knowledge. This approach consists of asking questions about the reactions of organisms to a specified environmental situation, and more particularly to a specified situation in the *physical* environment. Bioclimatology, on the other hand, has come to refer to an experimental approach which seeks information on the kinds of environmental situations producing a specified organismic reaction.

To put the term biometeorology in perspective, then, against this background of overlapping terminology, let us emphasize two portions of the definition mentioned previously. First, the term refers to the study of *interactions* in the sense that the direction of controlling processes may be from the physical environment to the organism, from the organism to the physical environment, or from both, depending upon the particular circumstances under consideration. Second, the term refers to scales of time of the order of a few hours or a few days.

The Structure of Biometeorology

Given our concern here with both biological and meteorological processes, it will be helpful to consider several ways of classifying specific approaches or frames of reference within each of these broad terms. For example, time and space criteria applied to the knowledge of the atmosphere might yield a system of approaches such as the following:

Micrometeorology—behavior of the atmosphere near the ground, primarily during short periods of time.

Aeronomy—similar to micrometeorology but restricted to the atmosphere well above the earth; related to the physical aspects of aerobiology.

Synoptic meteorology—although the term means the study of the atmosphere at an instant, it refers to the general field of short-term forecasting.

Climatology—study of the generalizations obtainable from comparisons among past examples of atmospheric behavior.

Alternatively, if one were to divide meteorology according to the type of biological regime in which knowledge is to be applied, the result might be:

Agricultural meteorology—atmospheric interactions with plants and animals raised primarily to feed man.

Forest meteorology—atmospheric interactions with plants and animals associated with lands producing wood fiber.

Air pollution meteorology—study of the results of interactions among sources of air contaminants, the dispersing atmosphere, and the receptors of the contaminants.

Limnological meteorology—study of the atmosphere as it affects fresh water biological systems.

Architectural meteorology—study of the interaction of the atmosphere with the shelters of man and animals, and so on.

Division of biological processes according to approach might result in a classification by type of organism, the broadest breakdown for our purposes being plants, animals, and humans. By function, the processes might be divided as follows:

Physiological—study of internal mechanisms maintaining life.

Ecological—study of the links between individual organisms and groups of organisms and their common environment.

Pathological—processes associated with disease or poor health.

Demographical—processes determining population size, density, and composition.

Biometeorology as a Science

Biometeorology is a science in several senses. First, it has developed to the point where its practitioners work with biological and meteorological concepts and data to produce valid and useful predictions. Many of these are mentioned later in this book, but it should be said at the outset that many of the techniques and procedures were developed long before the word "biometeorology" was even coined.

Mention of the relative newness of biometeorology prompts mention of its interdisciplinary nature. The ranks of scientists who call themselves biometeorologists contain individuals with a wide diversity of training and expertise, from medical scientists trained as medical practitioners, to plant physiologists, animal behaviorists, and dynamic meteorologists. At the heart of most biometeorological research is the instrumentation specialist, to whose work this book makes only passing references.

Finally, biometeorology is a science in the sense that scientists have organized themselves into various societies and groups with the express purpose of furthering the progress of their science and adhering to commonly recognized goals. The International Society of Biometeorology was founded in 1956 and has since held four international congresses, the most recent at this writing at New Brunswick, New Jersey, in August 1966. Many scientific societies have held technical sessions on biometeorology at their meetings: in the United States some of them are the Americal Meteorological Society, the Ecological Society of America, and Americal Physiological Society, and the American Society of Agronomy.

Because of its interdisciplinary nature, biometeorology requires of its practitioners a relatively wide variety of background knowledge. The biologists needs to know physics and meteorology, and the meteorologist needs to know some biology. They both need to know various mathematical and instrumental techniques. This need to know extends even to research teams where theoretically each scientist is a specialist in one area of biometeorology. With this introduction, the reader may move on to an exposition, and then a synthesis, of what was just referred to as the "background knowledge" underlying biometeorology.

SECTION I

The Physical Environment

Chapter 2

ENERGY AND ECOLOGY

The Basic Link in an Ecosystem

IN CHAPTER 1, it was emphasized that biometeorology is concerned with the study of interactions among biological and physical elements of certain systems. It seems clear that physical systems may exist without biological elements, but the converse is not true. What properties of the physical environment, then, underlie the evolution and maintenance of life? What are the processes linking these environmental properties to the organisms which represent life?

In answer to the question, a productive point of view is that the energy content of the physical environment underlies life and that various processes of energy transfer link the physical and biological elements of an ecosystem.

The Meteorologist's Role

Biometeorology is concerned with the transfer of energy between the organisms and the atmospheric part of the physical environment. Biometeorologists work to identify which aspects of life are responses to the atmospheric environment and to explain the reasons for the responses identified.

As suggested previously, there are several classifications of biometeorological effort and several distinct plans for research. Common to all, however, are a clear knowledge of the structure and behavior of the separate physical and biological elements and an understanding of the processes governing the transfer of energy among the elements.

The Biologist's Goal

The ecologist is faced with the task of explaining why species and individuals live where they do in the forms they do. To accomplish his task, he needs an adequate description of the physical environment of each organism and of the variability of the environment. He needs an understanding of the processes determining the range of environments in which the life of each species can be maintained.

The agronomist and the forester have the same need as the ecologist to know the mechanisms by which plants respond to the physical environment, though their objective is not so much to explain limits as to optimize plant response as expressed in some special measure of performance. Similarly, the animal husbandman wants to optimize response of a different set of biological processes to the same set of physical processes. Though the problems of the entomologist, the plant pathologist, or the environmental health physician may seem entirely different, all are concerned with the response of organisms to the physical environment.

Energy and Environment

Energy is the word we use in referring to that property of matter which may be transferred from one place to another, may appear in the form of motion or the potential for motion, and may appear in a form related to heat. From beginning physics, recall that positional energy is called potential energy, and the energy of gross motion is called kinetic energy. The internal energy of composition is chemical energy, and the internal energy of motion is heat.

Within the purely physical environment, energy exists in these different forms. It is converted and transferred by different processes. In the hydrosphere and in the lithosphere, the relative importance of these processes may be different from that in the atmosphere, but the processes are the same and are governed by the same physical principles. Thus, though biometeorology deals primarily with the atmospheric part of the ecosphere, and though the biologist's interest in it is centered on organismic responses to atmospheric variables, knowledge of energy transfer serves in the study of all parts of the environment.

Heat Transfer Processes

Heat is a form of energy. It defines in a general way the aggregate internal energy of motion of the atoms and molecules of a body. To cast concepts of heat transfer in a direct and fundamental framework of common experience, consider a small charcoal stove, with briquettes aglow, and resting on a table. A bare hand touching the stove on any side, top, or bottom, will have heat transferred to it in such great quantity and so fast by *conduction* that burning or at least discomfort will result at once. One removes his hand abruptly from contact with the stove. If the hand is removed to a point above the stove, the hot air currents rising from the stove may transfer heat to the burned hand so rapidly by *convection*, that acute discomfort may continue even though direct contact has been broken.

Removing the hand to a point some distance to the side of the stove will still not completely reduce the flow of heat below the point of discomfort to the hand, for it will continue to be transferred by *radiation*. While radiation accounted for some of the transfer to the hand held above the stove, its contribution in that case was probably overshadowed by that of convection. Conversely, if a small electric fan is blowing air past the stove to the hand held beside the stove, it may well be that the amount of heat being transferred by this horizontal convection, which meteorologists call *advection*, exceeds that transferred by radiation.

If on a second test, the hand is covered with a wet cloth glove before being touched to the stove, a great deal of heat will be conducted to the glove, but a relatively long time will elapse before one feels the pain. Since the same glove dry would not delay the pain nearly so long, and since the same amount of heat would be conducted in a given time, some other mode of heat transfer would be operating in the wet glove. Of course, the heat is transferred to the water, first raising its temperature and then evaporating it without further rise in temperature. So great is the amount of heat involved in the *latent heat* of vaporization in between stove and hand, very little is conducted on to the hand. This fourth mode of heat transfer is often overlooked by the casual student of these matters, but its great importance must be remembered in almost all studies of biometeorology.

Conclusion

In summary, the energy content of the physical environment supports life as energy is transferred to and from the organisms. Heat is the energy form of most direct interest in biometeorology, and it is transferred by the processes of radiation, conduction, convection-advection, and the movement and changes of the state of water. It is now our purpose to examine each of these modes of heat transfer as it acts in the natural environment. Following that, we will examine them collectively. In later sections, we will examine some of the myriad biological results of energy in the environment.

Chapter 3

RADIATION

ANY PORTION OF MATTER which is not at the absolute zero of temperature gives off energy to the surrounding space. The energy transferred in this way and the propagation process are both called radiation, and the process is most commonly and conveniently thought of as involving wavelike motion. Radiation is distinctly different from other modes of heat transfer in its great speed of propagation, which equals that of light, and because no material medium is required to exist across the path of transmission.

Spectra

In the model of wavelike propagation, the frequency ν is related to the wavelength λ by the formula

$$\lambda\nu = c$$

where c is the speed of light. Different radiation phenomena are classified according to wavelength, and the groups of wavelengths associated with the different phenomena are called the spectra of those phenomena.

Spectra may be continuous, so that all possible wavelengths in a given interval are involved in the radiation. They may also be present in bands, where several characteristic gaps appear in an otherwise continuous spectrum. Finally, line spectra are composed of characteristic groups of individual wavelengths. Most natural materials have radiation spectra appearing partly as bands and partly as lines. In many, some of the bands are so broad as to appear nearly continuous over wide regions of the electromagnetic spectrum.

Black Body

A black body is a material which has a continuous radiation spectrum in all wavelengths and which absorbs all radiant energy which falls upon it. As we shall see presently, these two properties are closely related. No known natural material is a perfect black body, although many are nearly so over large portions of their spectra.

Radiation Terminology

It will be helpful in understanding the concepts to follow to have well in hand several terms often used in discussions of radiation phenomena. First of all, there are three terms which are used to describe the passive reactions of materials to impinging radiation:

Absorptivity, a, is the fraction of incident radiation which is absorbed by the material.

Reflectivity, r, is the fraction of incident radiation which is reflected by the material.

Transmissivity, t, is the fraction of incident radiation which is transmitted by the material.

The values of these three coefficients for a given material are all numbers between zero and one, and they must add to one since all incident radiation must either be absorbed, reflected, or transmitted. The terms may be used to refer to the reaction of a material to individual wavelengths, to bands of wavelengths, or to the entire spectrum. The reflectivity of a material with respect to the individual wavelength λ, for example, would be written r_λ; while with respect to the entire spectrum it would be simply r.

The term albedo is used to refer to the reflectivity of a material summed over all wavelengths for a sunlit surface. Thus, albedo refers not only to all incident wavelengths but also to a particular quality of incident radiant energy—that of sunlight.

As noted, the terms just defined all pertain to the passive reaction of a material to incident radiation, and their values are all relative numbers whose base of comparison is the amount of incident energy. Since by definition, the absorptivity of a black body is 1.0, the values of r and t must be equal to zero.

Emissivity, ε, is a coefficient referring to the active behavior of a material with respect to radiation. Emissivity is the fraction of energy emitted by a material using the amount of energy emitted by a black body as the base of comparison. Again, one may use the coefficient to describe behavior with respect to individual wavelengths, bands, or the entire spectrum. By definition, the emissivity of a black body is equal to 1.0, with respect to wavelengths, bands, or the entire spectrum. That is to say, since ε is also always a number between zero and one, a black body emits the maximum amount of energy it is possible for any material to emit under the prevailing conditions of emission such as temperature.

Planck's Law

The first law of radiation we will need to discuss is Planck's law. The law may be stated in the form of an equation which gives the distribution of energy emitted from a black body as a function of wavelength and of the temperature of the black body:

$$E_\lambda = c_1 \lambda^{-5}[\exp(c_2/\lambda T) - 1]^{-1} \tag{3-1}$$

where E_λ is the amount of energy emitted in the band from λ to $(\lambda + d\lambda)$, T is the temperature in Kelvin degrees of the black body, and c_1 and c_2 are constants.* Because of the relationship of

* Values and units of the constants in Planck's law may be obtained from the following:

$$c_1 = 2\pi hc^2, \qquad c_2 = hc/k,$$
$$c_3 = 2\pi h/c^2, \qquad c_4 = h/k,$$

where

h is Planck's constant $= 6.55 \times 10^{-27}$ erg sec,
k is Boltzmann's constant $= 1.37 \times 10^{-16}$ erg deg^{-1}, and
c is the speed of light $= 3 \times 10^{10}$ cm sec^{-1}.

wavelength and frequency, Planck's law may also be stated as

$$E_\nu = c_3 \nu^3[\exp(c_4 \nu/T) - 1]^{-1} \tag{3-2}$$

where c_3 and c_4 are also constants.* It should be noted also that the connection between the two forms of the law is

$$E_\nu = \lambda^2 E_\lambda / c \tag{3-3}$$

where again c is the speed of light. E_ν is the energy emitted by the black body in the band from ν to $(\nu + d\nu)$, which as we shall see below is a bandwidth different from $d\lambda$, depending on the value of λ. Table 3-1 gives values of E_λ and E_ν for two values of temperature and several values of λ, according to the above formulas.

Table 3-1

BLACK BODY RADIANT FLUX[a]

T (°K)	λ (microns)	E_λ (ergs cm^{-2} sec^{-1} cm^{-1})	E_ν (ergs cm^{-2})
6000	0.25	2.67×10^{14}	0.05×10^{-4}
	0.55	9.66×10^{14}	0.97×10^{-4}
	1.0	3.83×10^{14}	1.24×10^{-4}
	1.5	1.24×10^{14}	0.93×10^{-4}
	2.0	0.50×10^{14}	0.67×10^{-4}
	3.0	0.13×10^{14}	0.38×10^{-4}
270	3.0	0.03×10^{7}	0.01×10^{-10}
	5.0	2.88×10^{7}	2.40×10^{-10}
	10.0	18.35×10^{7}	61.18×10^{-10}
	20.0	8.74×10^{7}	116.55×10^{-10}
	60.0	0.33×10^{7}	40.14×10^{-10}

[a] Data are for temperatures of 6000 and 270°K at representative wavelengths, expressed per unit wavelength and per unit wavenumber.

The distribution of energy from a black body according to Planck's law is usually presented as a graph of E_λ vs. λ. From Table 3-1 we see a plot for either temperature would show, for increasing λ, a rapid increase to a maximum value, then a decrease in E_λ tailing slowly off to lower values. However, the maxima would not occur at the same value of λ by any means. For $T = 6000$, λ_{max} is near 0.5 micron and for $T = 270$, it is near 10 microns. From Table 3-1, we may see also that, at the value of λ in common for the two temperatures as tabulated, the energy from the black body having the higher temperature is 10^8 or so greater than that from the cooler. This huge difference in scale makes it difficult t opresent on a single graph black body emission spectra from two bodies at greatly different temperatures. Such a plot may be shown as in Fig. 3-1 if only qualitative relationships are required

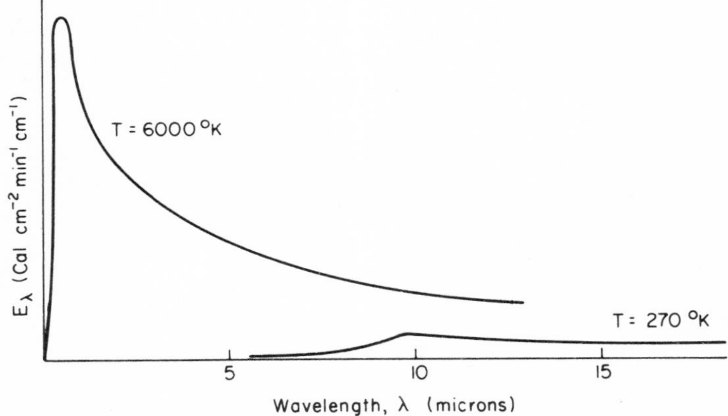

Fig. 3-1. Schematic representation of black body emission spectra, as a function of wavelength, for temperatures of 6000 and 270°K.

on the ordinate. In this figure may be seen the characteristic shape of a black body emission spectrum, which embodies the first radiation law: Planck's. The relative shape is the same regardless of the temperature: a single maximum with energy values tailing off toward longer wavelengths.

Although most references have presented black body spectra on a wavelength basis as in Fig. 3-1, there has recently been a good case made for presenting such spectra on a wavenumber basis according to the alternate form of Planck's law.[(1)] In Fig. 3-2 are presented the two forms of the plot side by side. One is on a linear

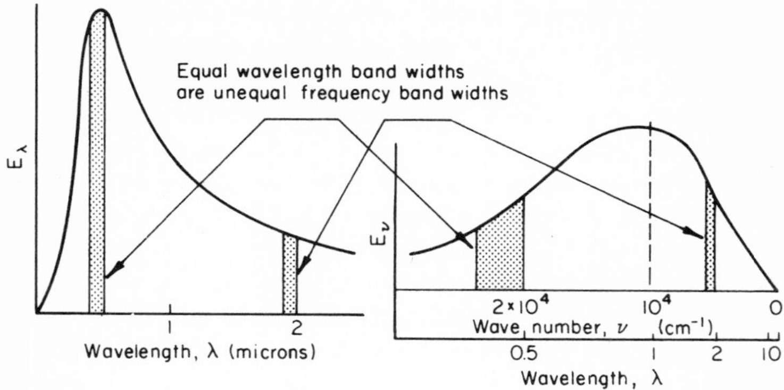

Fig. 3-2. Schematic representation of black body emission spectra for 6000°K as functions of wavelength and of wavenumber.

scale of wavelength, the other on a linear scale of wavenumber ($1/\lambda$) and thus a nonlinear scale of wavelength. Larger values of λ are toward the right in both cases. In either plot, the ordinate value is the amount of energy being emitted from a unit area of the black body (e.g., cm^2 as in Table 3-1) in a unit of time (e.g., sec) and in a unit wavelength or frequency band (e.g., $d\lambda$ cm). But because of the relationship between λ and v, a bandwidth of $d\lambda$ cm in one part of the spectrum has a different bandwidth in wavenumbers from that in another part of the spectrum. For example, let $d\lambda = 0.01$ micron, so that for $\lambda = 0.50$ micron the corresponding bandwidth in wavenumbers is from 1/0.50 to 1/0.51 micron, or from 20,000 cm^{-1} to 19,600 cm^{-1}: 400 units. For $\lambda = 1.00$ micron, the corresponding bandwidth in wavenumbers is from 10,000 to 9900 cm^{-1} and for $\lambda = 2.0$ microns, from 5000 to 4975 cm^{-1}: 100 and 25 units, respectively. Equal bandwidths in one mode of expression have unequal bandwidths in the other mode of expression.

Why even be concerned with two modes of expression? On the one hand we have most texts and references using wavelength, probably because wavelength can be measured and frequency can, as yet, only be calculated. On the other hand, we have the necessity for using frequency if the plotted spectrum is to be considered a true energy diagram: one in which a unit of area represents the same amount of energy in all parts of the diagram. Frequency, or wavenumber, makes the black body spectrum a true energy diagram because of another law due to Planck which says that the energy in a photon is proportional to its frequency, not its wavelength: $e = h\nu$, where h is known as Planck's constant. Thus, whenever the discussion of radiation requires a true energy diagram, one should employ a frequency scale.

Although most of the following discussion of radiation will be conducted in terms of wavelengths, the reader should understand clearly the differences between the two modes of expression and should note that in the transformation from a wavelength to a wavenumber, the location of the spectral maximum shifts. In Fig. 3-2, for a black body of temperature 6000°K, for example, the maximum is shifted from about 0.5 micron to about 1.0 micron. Since 6000°K is approximately the effective temperature of the direct beam of sunlight reaching earth, we see that the major portion of the energy of that beam lies in the part of the spectrum near 1.0 micron. One of the important consequences of this fact will be discussed later with respect to the heat load on plant materials.

As noted just above, the larger of the two temperatures in Table 3-1 and Fig. 3-1 is close to that of the sun, which for purposes of our general discussions may be considered nearly a black body. The lower of the two temperatures, 270°K, is representative of the mean temperature of earth, which for our purposes may also be considered nearly a black body.

In Fig. 3-1, one may see that all but about 1% of the energy from a black body with $T = 6000$°K lies between the wavelengths 0.15 and 4.0 microns, while for the cooler black body with $T = 270$°K, the same range is 4.0 to 80 microns. Although in the part of the spectrum near 3 or 4 microns we may see from Table 3-1 that in absolute energy the higher of the two temperatures produces a vastly greater flux, still when we consider energy in proportional units as above, we may for all practical purposes say that the dividing line between what we call *solar*, or shortwave, radiation and *terrestrial*, or longwave, radiation is 4.0 microns.

In most biometeorological treatments of heat transfer by radiation, it will be possible to deal only with these two bands. Yet despite this convenient artificial separation of radiation into these two bands, in some cases it would be necessary to recall that the two actually overlap. Common experience makes it rather easy to accept the fact that there is a good deal of longwave (heat) energy in sunshine, but less easy to accept the fact that the earth emits energy in the visible spectrum near 0.5 micron, though in amounts so small as to be undetectable by our eyes.

Wien's Law

The generalized statement of Wien's law may be seen in Fig. 3-1. It says that there is an inverse proportion between the black body temperature and the wavelength of its maximum emission: $T\lambda_{max}$ is a constant depending upon the units used for temperature and wavelength. Thus, in Fig. 3-1 for $T = 6000$, the λ_{max} is about 0.5 and for $T = 270$, the λ_{max} is about 10.0.

The Stephan–Boltzmann Law

In Fig. 3-1 may also be seen the essence of the Stephan–Boltzmann law: The total amount of energy emitted by a black body integrated over all wavelengths is proportional to the fourth power of its absolute temperature; total flux is σT^4, where the constant of proportionality σ is known as the Stephan–Boltzmann constant.* Thus, if we double the absolute temperature of a black body, we will increase the total rate at which energy is emitted by 2^4 or 16 times. In Fig. 3-1, the total flux is depicted by the total area under a particular curve, and even in this qualitative presentation the vastly greater flux from a unit area of the sun can be seen in comparison with that from a unit area of the earth.

A corollary of the Stephan–Boltzmann law is that a body cools or warms by radiation at a rate proportional to the difference between the fourth power of its own temperature and the fourth power of the mean temperature of its surroundings.

* The value of σ is 0.817×10^{-10} cal cm^{-2} min^{-1} deg^{-4}.

In summary of the first three radiation laws discussed, then, we see that Planck's law gives us the basic shape of the black body emission spectrum; Wien's law tells us the maximum in the spectrum shifts to longer wavelengths for lower temperatures; and the Stephan–Boltzmann law tells us that the areas under curves for various temperatures are proportional to the fourth powers of those temperatures.

Kirchhoff's Law

In the definition of a black body, the ideas of emission and absorption were seen to be connected. In fact, they are intimately and crucially connected through Kirchhoff's law: For a given wavelength and a given temperature, the absorptivity of a material equals its emissivity, or $a_\lambda = \varepsilon_\lambda$.

The importance of Kirchhoff's law may be seen through some simple examples. In Fig. 3-3 is a much enlarged portion of Fig. 3-1,

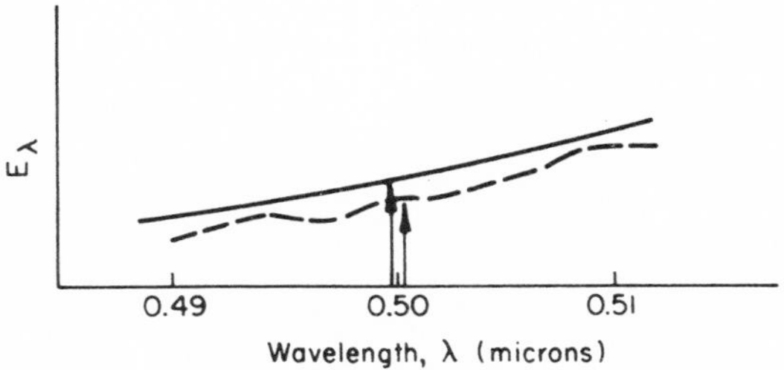

Fig. 3-3. Schematic representation of emission spectra of black body and of typical earth materials at 270°K and in the vicinity of 0.5 micron wavelength.

the part of the emission spectrum of a black body at 270°K in the region of 0.5 micron. Shown also is the emission spectrum for a typical material of the earth's surface. Both the black body and the earth will emit very small amounts of energy at 0.5 micron on an absolute scale, as seen also in Fig. 3-1. But by Kirchhoff's law, both are seen to emit very efficiently on a relative scale. To illustrate this with a simple and hypothetical numerical example, let us say that a black body at 270°K emits 10 units at 0.5 micron, and the earth's surface emits 9 units at the same temperature and wavelength. By definition, then, the emissivity of the earth is 0.9 at 270°K and 0.5 micron. We know that at 0.5 micron a large amount of energy arrives at the earth's surface in the solar beam, let us say 10,000 units in our numerical example. Since the earth's absorptivity here is equal to its emissivity, it will absorb 9000 units. Thus, even though earth emits almost no energy at the wavelengths near the sun's maximum on an absolute basis, it is still a very efficient absorber of the sun's energy at these same wavelengths, reflecting Kirchhoff's law. If this were not so, the earth's temperature would be much lower.

In Fig. 3-4 showing the simplified absorption spectrum of

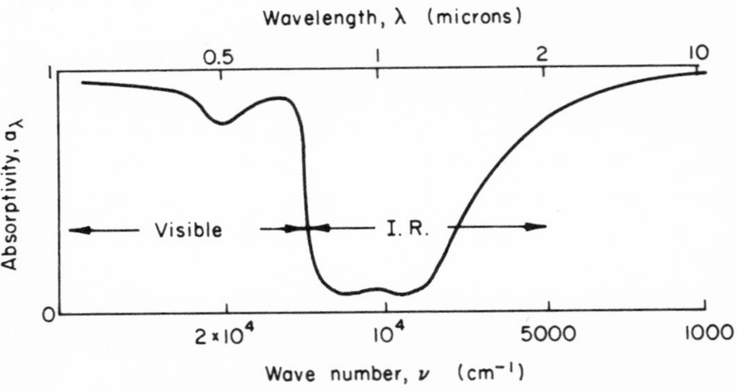

Fig. 3-4. Generalized representation of absorptivity for typical leaf materials, as a function of wavelength and of wavenumber.

typical leaf materials, Kirchhoff's law may be seen to operate in several most remarkable ways. First, notice that the relative minimum of absorptivity in the range of the visible spectrum lies at the wavelengths of green light. Therefore, since it is mainly in the green that visible light is reflected and transmitted (*r* and *t* are both nearly zero at other wavelengths), leaves appear green from above and below when illuminated by visible light. Notice second that the broad minimum of absorptivity in the region of infrared radiation from about 0.7 and 2.0 microns enables the leaf to escape the tremendous potential heat load at those wavelengths contained in the solar beam (see the wavenumber plot in Fig. 3-2). Finally, notice that in the long wavelengths near 10.0 microns, leaves have a very high absorptivity and emissivity. Because solar radiation contains only relatively small amounts of energy at this wavelength, the leaves, though absorbing efficiently, are not subjected to a major heat load. Since it is in the region of 10.0 microns and beyond that a material of terrestrial temperatures has its λ_{max}, however, leaves are able to dissipate their heat load very effectively by radiating in longer wavelengths. Thus, by Kirchhoff's law, an increase in the absorptivity of leaves in the infrared and a decrease near 10.0 microns would both result in marked increases in leaf temperatures. It may be noted in passing, without going into detail, that such changes in absorptivity in conjunction with changes in leaf size may result in changes in the heat load on a leaf as it develops from a newly emergent to a fully mature leaf.[(2)]

Earlier we noted that the terms absorptivity, reflectivity, and transmissivity may refer either to individual wavenumbers (or wavelengths) or to bands of wavenumbers. With information now at hand about Planck's law and about an absorption spectrum such as that for leaf materials, consider the physical meaning of, say, the effective absorptivity of leaves over the band from ν_1 to ν_2. Begin by recalling that the energy arriving at the leaf surface from all surrounding sources in the small band ν to $\nu + d\nu$ is of a form $E_\nu' \, d\nu$. The primed symbol is used to indicate that the sources need not be black bodies.

Of the incident radiation in the band ν to $\nu + d\nu$, the amount absorbed is $a_\nu E_\nu' \, d\nu$, and over the larger band of interest, the total energy absorbed is the integral

$$\int_{\nu_1}^{\nu_2} a_\nu E_\nu' \, d\nu.$$

Since over ν_1 to ν_2 the total incident radiation is the same integral without the a_ν, the effective absorptivity for the band is, by the definition of absorptivity, the ratio

$$a_{\nu_1-\nu_2} = \int_{\nu_1}^{\nu_2} a_\nu E_\nu' \, d\nu \bigg/ \int_{\nu_1}^{\nu_2} E_\nu' \, d\nu. \qquad \textbf{(3-4)}$$

As will be mentioned in Chapter 8, when ν_1 and ν_2 span the short-wave spectrum, the effective absorptivity of leaves, obtained by using measurements in Eq. (3-4), is near 0.50.

Spectra of Principal Atmospheric Gases

Figure 3-5 presents in a very simplified form the absorption spectra of water vapor and carbon dioxide, the two principal atmospheric gases so far as environmental radiation is concerned. Note that the plot here is on a wavelength scale so as to expand

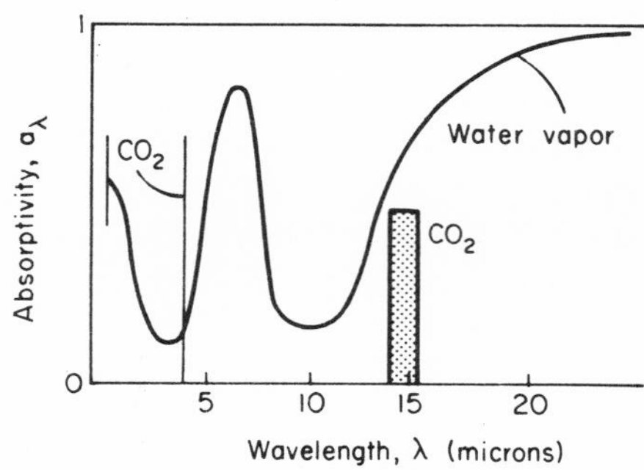

Fig. 3-5. Generalized representation of absorptivity for water vapor and carbon dioxide, as a function of wavelength.

the information for easy presentation. In this simplified form, water vapor is seen to have its principal absorption bands at 3, 5–7, and beyond 13 microns. Carbon dioxide absorbs most strongly at 4.3 and 14–15 microns. In the "short wave" spectrum—not shown in Fig. 3-5—these two gases have some narrow absorption bands whose total effect is a relatively small absorptivity for sunlight; hence, a cloudless sky is relatively transparent to sunlight. On the other hand, the atmosphere is not so transparent to terrestrial, or long wave, radiation. Even in this region, though, there is a band between 8 and 13 microns in which the atmosphere is transparent; this band is the so-called "window" in the atmosphere.

In the window, energy passes almost unimpeded from the sun

and sky toward earth and from the earth toward space. The energy in other wavelengths—which is by far the greater portion—is soon absorbed by the atmosphere upon leaving the earth's surface. The net effect of the unimpeded arrival of sunlight on earth and the greatly impeded flow of radiant energy outward from earth is a retention of energy within the earth-atmosphere system. This net effect of differential absorption is called the "greenhouse effect." The name derives, of course, from the very similar differential absorption of glass in an ordinary greenhouse. The glass acts as do the atmospheric gases: passing most shortwave radiation but only certain wavelengths of longwave radiation. It should be noted that the window lies near the part of the longwave spectrum with nearly maximum emission by the earth.

Energy exchange by radiation takes place within the earth-atmosphere system in a very complex manner in the wavelengths outside the window. Figure 3-6 depicts this exchange. The earth

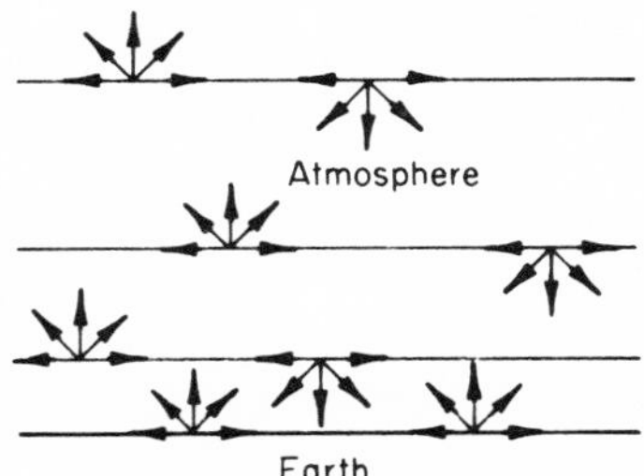

Fig. 3-6. Schematic representation of longwave energy transfer within the earth-atmosphere system.

radiates energy to the water vapor and carbon dioxide of the air, and vice versa. Adjacent layers of air exchange radiant energy with one another in these same wavelengths outside the window. Absorption is so efficient in these wavelengths that all of such energy leaving the earth or one of the atmospheric layers is completely absorbed within a matter of meters from its origin. The length of this path is shortened by greater amounts of water vapor in the air and even more drastically when the water vapor is condensed to a liquid.

The direction of net effect of radiant energy exchange in opaque wavelengths between any two layers depends on the temperatures of the layers compared with each other and with that of the earth's surface beneath, the warm layers losing heat to the cooler. On a clear and dry night, the window is "wide open" and heat which is lost directly to space from earth in the wavelengths 8–13 microns is replaced by heat being radiated to the surface from the lowest levels of air in the opaque wavelengths. The surface soon becomes colder than nearby layers of air, and the layers of air nearest the surface become in turn colder than the layers next above them. The results are different, however, when there is a layer of clouds, as we shall see presently.

Unlike water vapor, liquid water has a high albedo and a comparatively high absorptivity in the wavelengths of the window. The results are that a cloud layer reflects most sunlight from its top and absorbs most longwave radiation from below. With clouds present, as our common experience tells us, much less solar energy reaches earth. Thus daytime maximum temperatures are reduced when clouds are present. This results from liquid water's high reflectivity in short wavelengths. Its high absorptivity in long wavelengths prevents energy from escaping directly to space from the earth's surface. Instead, the surface "sees" the temperature of the clouds and tends toward that temperature as a minimum rather than toward the much lower temperatures of space. On balance, then, clouds act thermostatically to reduce the range of temperatures between day and night at the surface.

Energy Balance of the Earth-Atmosphere System

The net effects of all these radiation processes, and other heat transfer processes to be studied in greater detail later, are summed up in Fig. 3-7 as an energy balance of the earth-atmosphere system. The values in Fig. 3-7 represent conditions averaged over all latitudes and times of year. Conditions at any one place and time might well be quite different insofar as the values attached to each process are concerned, but all the processes would be active and the energy balance at least as complex as is shown.

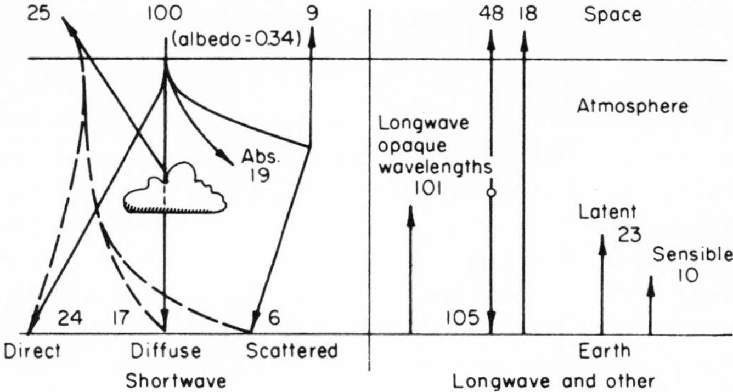

Fig. 3-7. Relative net transfer of heat within the earth-atmosphere-space system.

The arbitrary unit for comparison in Fig. 3-7 is chosen so that 100 such units enter the atmosphere from the sun.* The diagram is divided so as to represent three levels of the system—space, the atmosphere, and the earth—and two kinds of radiant energy—shortwave on the left and longwave on the right. Of the 100 entering units, 34 are returned as shortwave radiation from the earth and atmosphere: Therefore, we say the albedo of the earth-atmosphere system is 0.34, or 34%. Part of the energy reflected from the earth-atmosphere system comes from the scattering due to the materials of the atmosphere: gas molecules and solids. Part comes from reflection of cloud tops. The remainder comes from reflection at the earth's surface, some having traveled through a cloud layer and some not.

Of the light energy reaching the earth's surface, 47 units (24 + 17 + 6) are absorbed. Since the mean albedo of the earth's surface is about 0.10, the 47 units must represent 90% of the total shortwave arriving at the surface. About 5 units have been reflected and appear as part of the 25 reflected from the earth and from clouds together. Clearly, since the albedo of the earth is 0.10 and

* The actual value of these arbitrary units is very close to 2.0 cal cm^{-2} min^{-1} on a surface normal to the solar beam, a value known as the "solar constant."

that of the earth and atmosphere together is 0.34, the albedo of cloud tops is a good deal greater than 0.34, and the 34% is a weighted average of the several albedos involved. Anyone who has flown above cloud tops has experienced the blinding reflection from the clouds and the comparatively dull, or nonreflective, appearance of most parts of the surface beneath.

In the shortwave portion of Fig. 3-7 we see 19 units entering from space as shortwave energy and absorbed within the atmosphere without ever reaching the earth (shortwave absorptivity of the atmosphere 0.19). Unlike the 34 units involved in reflection, these 19 units are "converted" to longwave energy before continuing their journey. In the longwave diagram to the right, these 19 units appear as part of the energy whose arrows originate within the atmosphere: 48 going from the atmosphere to space and 105 going from the atmosphere to earth.

One-hundred-and-five units? How can this be with only 100 units entering? The explanation for this and other apparently large numbers in the longwave diagram lies in the fact that the shortwave flux shown takes place almost exclusively on the sunlit side of the earth, while the longwave flux is taking place on all sides at all times, as in the case with the latent heat and convective

Table 3-2
BUDGET FOR THE EARTH'S SURFACE

Arriving:	
Direct sunlight	24 units
Diffuse sunlight (through clouds)	17
Scattered sunlight (blue sky)	6
Longwave flux from the atmosphere	105
	152
Leaving:	
Absorbed by the atmosphere in wavelengths outside the "window"	101 units
Directly to space in "window" wavelengths	18
Latent heat (evaporation)	23
Convection	10
	152

heat fluxes shown. To satisfy himself that the diagrams do indeed represent a balance, one may check to see that the sums associated with arrows whose heads terminate in any one of the three portions of the system are equal to the sums associated with arrows originating in that same portion. For example, the budget for the earth's surface is shown in Table 3-2.

In summary of Fig. 3-7, then, we may say simply that in the earth-atmosphere environment the great majority of the heat-energy transfer takes place by radiation, and this as the result of a complex interaction among the radiation laws discussed and along a great variety of paths. This complexity we will examine in further detail later.

REFERENCES

1. See e.g., *Science* p. 1239 (Dec. 3, 1965) and p. 400 (Jan. 28, 1966).
2. D. M. Gates, Spectral properties of plants, *Appl. Opt.* **4**, 11–20 (January 1965).

PROBLEMS

3-1. Calculate the proportion of the energy being radiated by a material of emissivity 0.90 and temperature 27.3°C relative to the energy being radiated from an equal area of another material of emissivity 0.50 and temperature 0°C.

3-2. Imagine two flat, parallel walls, infinite in extent, and separated by a vacuum gap. The walls and the air are at radiative and thermal equilibrium. Wall 1 is a black body. Wall 2 has zero transmissivity, but it is not a black body. Write the equation which says "the energy arriving at Wall 2 from Wall 1 equals the energy leaving Wall 2 toward Wall 1." By imposing a single condition on temperatures within this two-wall system, use your equation to prove Kirchhoff's law.

3-3. If the absorptivities of water or carbon dioxide or both were larger in the band of the "window," temperatures on earth would be higher. Explain why this is so. Describe a modification to a horticultural greenhouse which would be analogous to these changes in the atmospheric greenhouse. Would the modification have the same effect on air temperatures in the horticultural greenhouse?

3-4. By actual calculation, confirm the values of E_λ and E_ν associated with a wavelength of 3.0 microns in Table 3-1.

3-5. A radiation thermometer measures surface temperature by assuming the surface viewed is a black body, then converting the radiant energy flux it senses to a mean surface temperature by means of the Stephan–Boltzmann equation.
(a) If the two walls in Problem 3-2 are actually at a temperature of 180°F, and the radiation thermometer is pointed at Wall 2, what temperature will it register?
(b) If Wall 1 is actually at 40°F and Wall 2 actually at 180°F, but the radiation thermometer registers a temperature of 150°F for Wall 2, what is the emissivity of Wall 2?
(c) If Wall 1 is at absolute zero and Wall 2 at 180°F, what is the emissivity of Wall 2 if its indicated temperature is −40°F?

3-6. In the laboratory you have measured the absorptivities of a material at each of several specific wavelengths between 3 and 60 microns.
(a) If the absorptivities were measured while the material was at 270°K, calculate an estimate of the effective emissivity of the material for the waveband 3 to 60 microns at 270°K.

Wavelength (microns):	3	5	10	20	60
Measured absorptivity:	0.9	0.8	0.2	0.4	1.0

(b) Is the effective emissivity you have calculated equal to the effective absorptivity with respect to incident radiation arriving from a black body which has a temperature of 6000°K?

3-7. Assuming the following diameters and temperatures for sun and earth, assuming both are black bodies at their respective temperatures, and assuming they are 93 million miles apart, calculate the ratio:

$$\frac{\text{rate of arrival of direct solar radiation}}{\text{rate of longwave emission}}$$

for (a) the earth as a whole and (b) the earth-atmosphere system as a whole. How do your two calculated ratios compare with the ratios obtainable from values in Fig. 3-7?

	Sun	Earth–Atmosphere	Earth
Diameter (miles)	865,000	8,000	8,000
Temperature (°K)	6,000	270	300

3-8. Calculate the percentage increase in the energy emitted by a material of emissivity 0.80 when its temperature increases from 40 to 100°F if its emissivity does not change with this heating.

Chapter 4

ENVIRONMENTAL TEMPERATURE

Temperature versus Heat

TEMPERATURE IS A MEASURE of the mean kinetic energy per molecule (speed) of the molecules in an object, while heat is a measure of the total kinetic energy of all the molecules of that object. A large object may have a much lower temperature than a small object and still have a greater heat content by virtue of the larger number of molecules in it.

So long as we confine our attention to but one object, its temperature changes represent proportional changes in heat content. With objects representing a variety of masses and types of material, however, this equivalence of heat and temperature disappears. This point is particularly important in studies of energy flow in an environmental system. For example, not much energy is represented in the sun's heating a thin layer of needle litter in the forest to very high temperatures. If one were to scoop up several handfuls of these hot needles and place them on a nearby granite boulder, heat would at once flow from needles to boulder but the boulder would show scarcely any rise in temperature. In another example, the same granite in the form of the wall of a building of a city would represent a tremendous source of heat for adjacent air during the night, even though the wall might never have risen to a very high temperature during the preceding day.

An additional aspect of the difference between temperature and heat appears upon consideration of a wet field in the early morning sun. A great deal of heat is being transferred from the field to the air as water evaporates and rises as vapor; but the temperature of the surface of the field changes only slightly.

"True Air" versus "Test Body" Temperature

"True air" temperature, also called "free air" or "shelter" temperature, is that measured in some sort of standardized shelter in a prescribed way. It is intended to be a measure of the true mean speed of the air molecules in the vicinity. Because the introduction of the measuring instrument into the environment to be measured itself introduces errors into the measurement process, certain precautions are taken and these precautions are set in a standardizing procedure. For example, the mean speed of nearby molecules of air would be increased (the temperature would rise) if heat were absorbed by the thermometer and then transferred to the nearby air molecules. Hence, a shelter shielding the thermometer from the sun is part of the standardizing procedure, as is the ventilation of the thermometer shelter to keep the thermometer in good contact with the wider population of air molecules it is meant to represent. In highly specialized research, the same errors avoided by shielding are often avoided by making the thermometer probe —usually electrical—so small as not to absorb very much solar radiation.

In a great many research problems, the true air temperature is not what is needed in a temperature measurement. Rather, the investigator requires the temperature of a body—possibly representing an organism—which is exchanging energy with the air nearby and also with such sources of radiant energy as the sun. This is often called a "test body" temperature. It represents the true heat load on the body, and it includes the effects of some of the heat fluxes specifically suppressed in the case of the true air temperature measurement. In what follows, the expression "air temperature" will be used to refer to a measure only of the true air temperature. Most of the examples cited, which include many measurements taken very close to the ground surface, are from research results obtained with the very small probes mentioned above.

Air Temperature near the Ground: Time Variations

Figure 4-1 shows a typical plot of air temperature near the ground through the course of a clear, sunny day. Temperature values are omitted in order to concentrate on the time relationships involved. Examining two of these relationships will serve to introduce several ideas discussed later.

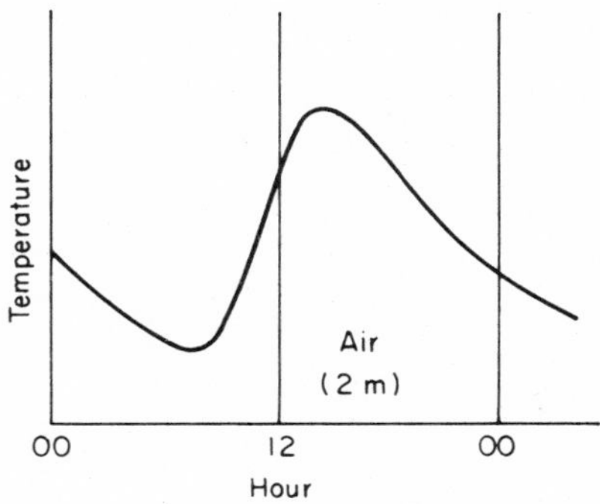

Fig. 4-1. Generalized diurnal trace of air temperature several meters above the ground.

First of all, note that the time of maximum air temperature lags that of solar noon, often by several hours under proper circumstances. The explanation is that, in the layer of air where this temperature is measured, energy is arriving at noon faster than it can be dissipated. This excess continues for a while even after solar noon—the time when the sun is highest in the sky on a given day—with the result that the temperature continues to rise. The rise ceases when the means for dissipation equals the arrival rate. The lag of air temperature after solar noon is said to be due to the "thermal inertia" of the soil-air system.

The other relationship to be noted is that the air temperature rises more rapidly in the morning than it falls during the afternoon and evening, resulting in an asymmetry. In the morning, the rate at which heat is being dissipated by evaporation and the rate at which the soil heat reservoir is being refilled are small compared with the rate of arrival of energy from the sun. The excess heat goes into heating the air. In the late afternoon and evening, rather than being a site for accumulation of heat, the soil heat reservoir supplies heat to the soil surface and thus also to the air near the surface. While this flow of soil heat is not large enough to keep the air temperature from falling, it retards the fall sufficiently to produce the observed asymmetry.

Detailed consideration of the interactions of these processes of energy dissipation will be postponed until later chapters.

Air Temperature near the Ground: Space Variations

Figure 4-2 shows how air temperature varies with distance from a bare soil surface at two different times of day. Such a presentation of the vertical distribution of a property (in this case

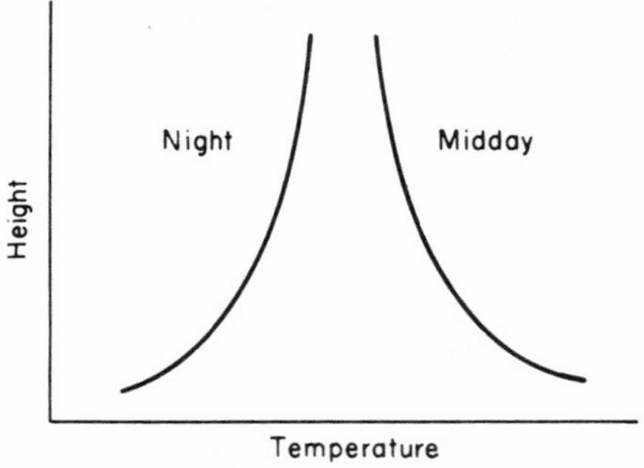

Fig. 4-2. Generalized temperature profiles near the ground at night and at midday.

temperature) is called a *profile* of that property. The curve with the higher temperatures is typical of midafternoon on a warm, clear day. The cooler curve is typical of late night hours. Two points are worth noting and will serve as an introduction to important ideas.

The first point is that both day and night temperature extremes are most likely to be found at the ground surface. The explanation is that this surface is the most active heat exchanger in the soil-air system. The second point is that, for a given difference in height, greater differences in temperature are found near the surface than farther from it. That is to say, the temperature *gradient* is greater near the ground both day and night. The explanation for this relationship also lies in the fact that the surface is the primary heat exchanger in the system.

Temperatures near the Ground Surface: Time and Space Variations

Combining the variations shown in Figs. 4-1 and 4-2, and extending the information to shallow depths in the soil, produces results such as those in Fig. 4-3. This presentation of temperatures

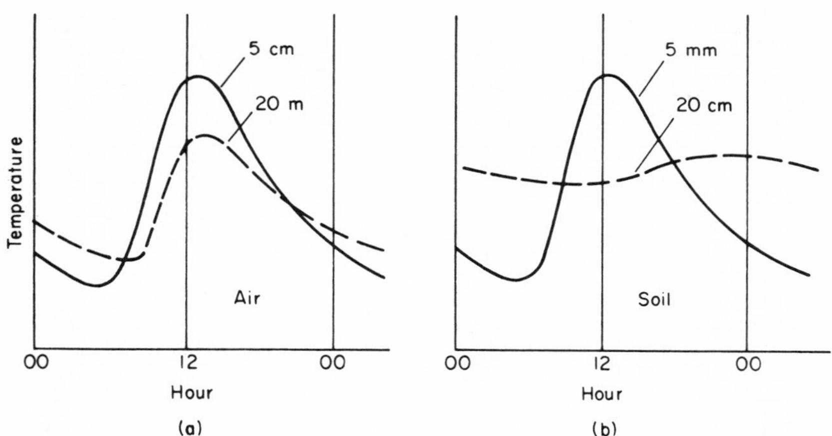

Fig. 4-3. Generalized diurnal traces of (a) air temperature and (b) soil temperature, each at two distances from the soil surface.

as time series is one of several commonly used to combine information on temperature, time, and distance in one diagram; and it serves to emphasize several more relationships. In both air and soil, there is a greater lag in the time of temperature maxima and minima at greater distances from the soil surface. Also in both air and soil, there is a greater reduction in the amplitude of the temperature wave at greater distances from the surface. Because air is fluid and can transfer heat by convection, and because soil is solid and transfers heat principally by conduction, the same relative effect reaches much higher into the air than down into the soil. For the same reason, the time lag is much less in air than in soil. The reduction in amplitude of the temperature wave and the lag time are much greater at only 20 cm in the soil than at 20 meters in the air. The physical laws governing these temperature differences and the properties characteristic of the two heat transfer media will be discussed in more detail presently.

What happens to these relationships under cloudy conditions or in winter? First of all, the mean temperatures in each level of the air or soil would be somewhat lower—the whole cycle would take place at lower temperatures. More than that, the amplitude of the temperature wave would be less at any given level. Finally, the time lag, or phase shift, between any two levels would be increased. These effects are all consequences of the reduced rate of input of solar heat at the surface and of the less effective, less vigorous convective and conductive linkages between two levels.

There are, in addition to the scheme in Fig. 4-3, two other commonly used presentations of the relationships among temperature, time, and distance. The curves of temperature vs. distance in Fig. 4-4 are called *tautochrones*, each line representing conditions

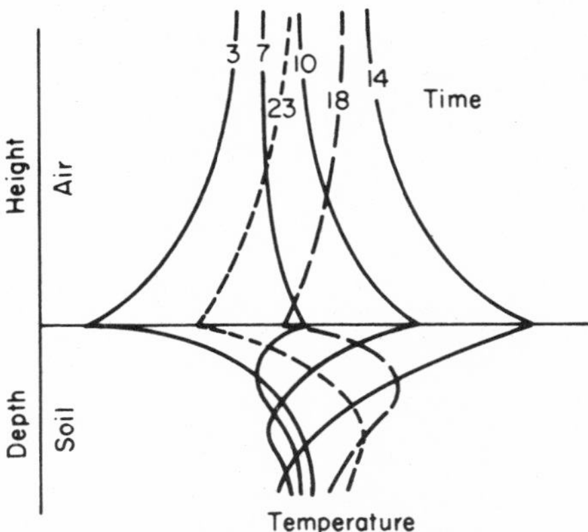

Fig. 4-4. Generalized soil-air temperature profiles (tautochrones) near the soil surface, for four-hour intervals during a diurnal period.

at a different time. Here may be seen the two curves of Fig. 4-2 lying at the extremes. Careful consideration of these tautochrones will reveal all of the relationships discussed above with respect to amplitude reduction and lag time. In Fig. 4-5, the same relationships are depicted in a plot of isotherms—lines of equal temperature—on time and distance coordinates.

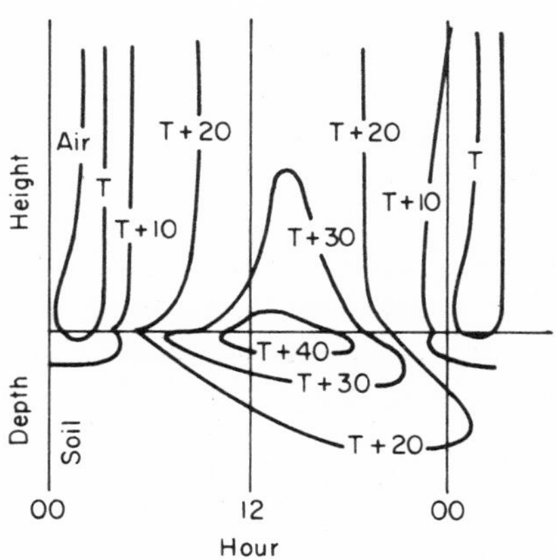

Fig. 4-5. Generalized diurnal patterns of isotherms near the soil surface on coordinates of time and distance.

Temperature Lapse Rates

Above, we have used the term gradient to refer to an amount of temperature difference over a given distance. This is an expression of the rate of temperature change with distance, and in referring to the vertical gradient in air the term *lapse rate* is used. This term is seldom used with reference to soil temperatures. In addition, the lapse rate of temperature in air is positive or negative depending on whether the temperature decreases at greater and greater heights or increases. The temperature lapse rate, then, is the rate at which temperature decreases with height. The unit employed is (temperature difference/height difference)—usually normalized to (°C/100 meters).

Several particular lapse rates have special physical significance and are therefore often used as bases for comparison with other values of the lapse rate. When there is no change of temperature with height, the lapse rate is *isothermal.* Within a well developed and growing cloud, the lapse rate of about +0.6°/100 meters is called *moist-adiabatic.* When air rises without condensation and cloud formation, it cools at the rate of 1.0°/100 meters. Thus, +1.0°/100 meters is called the *dry adiabatic* lapse rate and it is observed in any well-mixed layer of air found more than several hundred meters above the surface, provided no clouds are present. Nearer the earth's surface—in the range of several tens of meters—the lapse rate is often *autoconvective*: +3.4°/100 meters. Under these conditions, the air has constant density with height; hence, further heating in the lower layers and an increase in lapse rate would result in reduced density underlying a greater density, followed by overturning as the dense air moves beneath the less dense air to return the air column to stability. This overturning, or autoconvective motion, occurs only if the air is free to move. Very near the ground, the air does not have this freedom of motion, and during midday extremely large lapse rates may result.

All the lapse rates identified thus far have been positive—temperature has decreased with increasing height. The term "lapse" is used to refer to any condition in which a positive lapse rate exists. For conditions of negative lapse rate, the term "inversion" is used.

As suggested previously, more and more positive lapse rates occur with greater and greater heating of underlying layers of air and increasing tendency for overturning between layers. Lapse rates more positive than the autoconvective imply a condition in which overturning is actually suppressed or prevented. In the same manner, more and more negative lapse rates occur with less and less tendency for overturning. Under inversion conditions, the underlying air is more dense than that in layers above and the more intense the inversion, the greater external force would have to be applied to bring about overturning. Larger negative lapse rates simply represent more stable vertical stratification of air layers.

Lapse Rates above the Surface

In Fig. 4-6 appear frequency distributions of lapse rates observed hourly in three adjoining air layers throughout an entire year. Data from all hours are pooled: day and night, winter and

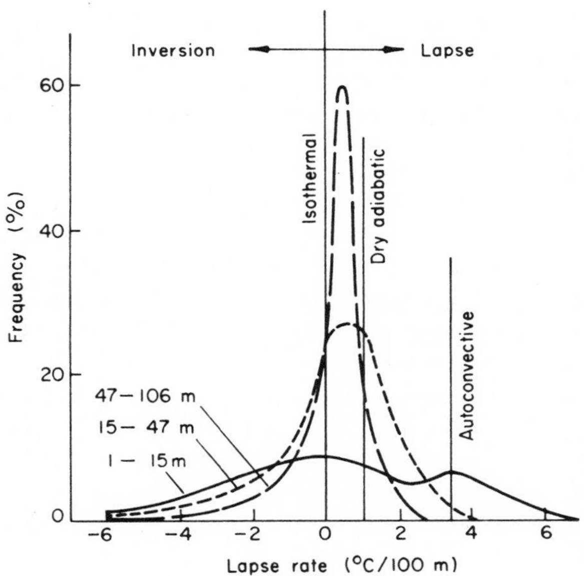

Fig. 4-6. Typical frequency distributions of lapse rate observed hourly through an entire year in three layers of the lower atmosphere. (Based on Fig. 30 of Geiger.(1))

summer. Several of the concepts just discussed are represented in these frequency distributions. To begin with the 59-meter-deep layer farthest from the surface, and thus least restricted in any tendency for overturning, lapse rates observed do not vary widely and are very much the most frequent between isothermal and dry adiabatic. As one would expect, in a layer which is free to move in the vertical, the characteristic lapse rates are those of a well-mixed layer—sometimes with and sometimes without condensation. The freedom for overturning is sufficiently great, notice, that in the upper layer autoconvective lapse rates were only approached but never observed. Inversions, on the other hand, were observed, and in some considerable strength. These were doubtless observed during calm and clear winter nights when inversions are deepest and best developed.

In the middle layer, 32 meters deep, the variety of lapse rates observed is much greater. The predominant values are again near those for well-mixed air layers, but there are now occasions when superheating and suppressed overturning result in observations exceeding the autoconvective lapse rate. Nearer the surface than the first layer discussed, this one takes part in more and stronger inversions.

Finally, the lowest layer observed—based at only 1 meter and only 14 meters thick—exhibits an even wider variety of conditions through the course of a year. Super-autoconvective lapse rates are almost commonplace, but the autoconvective rate itself has sufficient physical significance that a secondary mode of frequency occurs at about +3.4°/100 meters. Again, more and stronger inversions occur in this lowest layer. The broad mode of lapse rates near isothermal reflects the tendency for slight inversions to form just above the surface even during nights when upper layers may be mixed by moderate winds.

In summary of Fig. 4-6, several points bear repeating. In contrast to the "tailing off" of frequencies toward inversion, there is a rather definite upper limit cutting off the curves under lapse conditions. This, of course, is a reflection of the inherent stability of inversions and instability of extreme lapse conditions. The figure makes clear the striking increase in variety of thermal structure as one approaches the surface of the earth.

Lapse Rates in the Surface Layers

We shall now enter the realm very near indeed to the surface—one in which suppressed mixing results in even greater variety of lapse rates and extremes of temperature. To understand the dynamics of motion and transfer processes here, we divide the layer into four parts for purposes of description. As with the layers in Fig. 4-6, these will be thinner and thinner near the surface. In actuality, the layers are not really separable and do not always have the same characteristic vertical dimensions. They blend into one another and their dimensions depend upon time of day, time of year, and the nature of the surface materials—primarily the surface roughness.

The "boundary layer" rests on the surface and is only a matter of a few millimeters thick. It is truly the skin of air next to the surface. Here motion is restricted to the molecular scale, with the result that transfer processes are by pure conduction or molecular diffusion, with occasional formation of tiny and short-lived "chimneys" of very locally organized air motion.

In the "intermediate layer" is found rudimentary convection to the extent that organized motion takes place over short paths. But the paths are never as long as the few centimeters of the layer's depth; hence, transfer is never achieved by a mass of air moving through the layer. As a result, during midday for example, temperature of the air is never equalized across this layer despite wide fluctuations in temperatures at both the top and the bottom of the layer. It is the sheer physical presence of the solid surface presenting an obstacle to vertical motion which suppresses the development of convective motion. In Fig. 4-7a may be seen typical midday temperature traces made by very fast response instruments located near the top and the bottom of the intermediate layer. The highest temperatures at the top of the layer are never as high as the lowest in the lower level.

Convective eddying is more fully developed, and vertical transfer processes less suppressed in the several-meter-thick "overlayer" next above the intermediate layer. Although, as shown in Fig. 4-7b, there is a sizable lapse rate across the layer on the average, there are frequent cases when top and bottom temperatures are equalized. These indicate that the path length of the organized convective motion is sometimes as long as a few meters but is most frequently considerably less.

In the "macrolayer," convection is fully developed. In this layer several tens of meters thick, autoconvective motion is common and conditions are those of the lowest layer in Fig. 4-6. Here the effect of the earth's surface in suppressing vertical motion is fast disappearing, and the surface has its effects mainly on wind

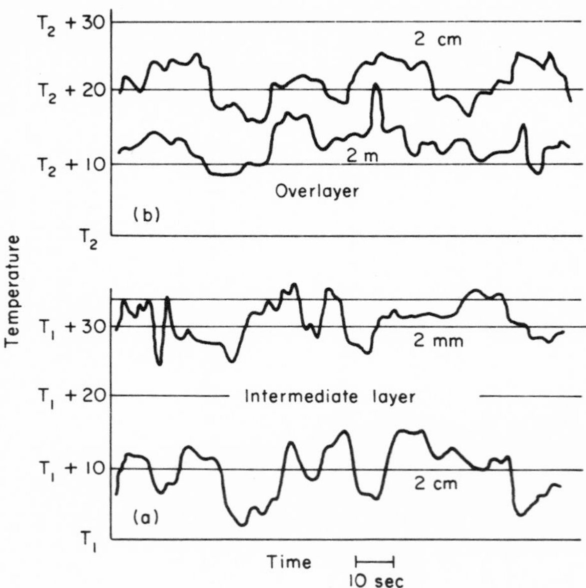

Fig. 4-7. Generalized temperature traces typical of those obtained with fast-response thermometers. (a) Traces for the bottom and top of the intermediate layer and (b) traces for the bottom and top of the overlayer.

speed and direction because of the frictional forces it exerts on the moving airstream.

Table 4-1 summarizes the discussion of the air layers immediately above the earth's surface. Again, the division into four layers is arbitrary and for descriptive purposes only. Actually the layers blend into one another, and so do their characteristic properties. In the table may be seen the enormity of temperature lapse rates near the surface—five orders of magnitude greater than those only a few meters above. Midday conditions are given.

Some Other Aspects of Air Temperature Variation near the Ground

With some idea of how temperature varies in time and space near the earth's surface, and a notion of the transfer processes in

Table 4-1
REPRESENTATIVE DIMENSIONS AND PROPERTIES INDICATING STRUCTURE OF THE AIR LAYER NEAREST THE EARTH'S SURFACE

Layer	Typical dimensions	Typical temperature difference across layer, and normalized lapse rate	Characteristics
Boundary	0 to 2 mm	25° (12.5×10^5)°/100 meters	Pure conduction, intermittent "chimneys" of air, convection absent
Intermediate	2 mm to 2 cm	15° (8×10^4)°/100 meters	Rudimentary convection, suppressed eddying, no mass exchange completely through the layer
Overlayer	2 cm to 2 meters	5° 250°/100 meters	Convective eddying more fully developed, occasional mass transfer through layer, autoconvective lapse rate at times near top of the layer
Macrolayer	2 to 50 meters	1.5° 3°/100 meters	Convection fully developed, autoconvective lapse rate common, influence of the ground mainly on wind speed and direction by friction

the lowest layers, we shall return to an examination of variations in and above the macrolayer, concentrating on transfer processes and their consequences as seen from several points of view.

Figure 4-8a shows, for an open and level environment, how the profile of temperature changes with time on a typical sunny morning. By now we are familiar with the increasing variability of temperature and extremes of lapse rate near the surface. This

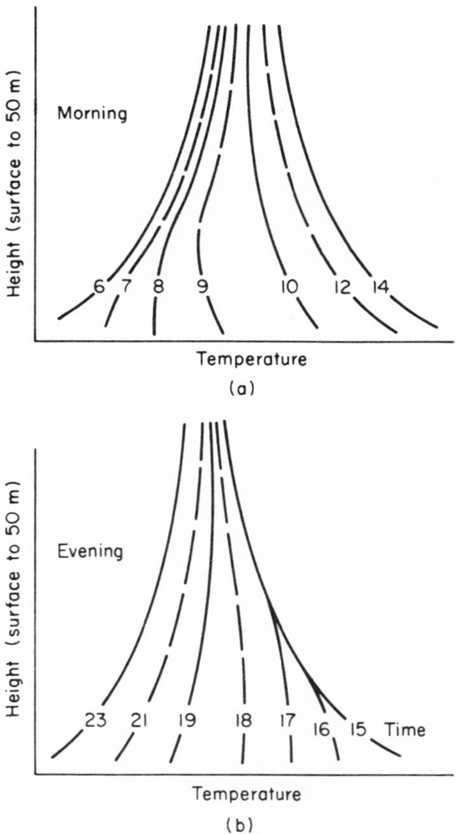

Fig. 4-8. Generalized profiles of air temperature near the soil surface in the hours (a) following sunrise and (b) following sunset.

results from the fact that the surface is the primary heat exchanger, which is in turn due to the fact that, at the surface, solar energy is absorbed and introduced into the soil-air system. At the surface, in the morning, bubbles of superheated air form. They grow in size depending upon such factors as wind speed and surface roughness, and then rise from the surface. As they rise, the bubbles shed some of their mass and heat at each level. They are replaced at the surface, according to the law of mass continuity, by subsiding air which has a higher heat content than the surface air between the "hot spots." The net result of the action of these processes of rising bubbles and subsidence is shown by the successive profiles in Fig. 4-8a. While inversion conditions are preserved at higher levels well into the morning, in the lowest 10 meters or so, inversion is replaced rather early by lapse rates approaching the autoconvective. By midmorning, inversion has disappeared from all levels.

As the sun approaches the horizon and the input of solar energy is reduced during the afternoon, temperature profiles change in a nearly mirror-image fashion compared with the changes during the morning. Figure 4-8b shows how this is so. Although the earth's surface is still acting as the primary heat exchanger in the system, the dominant mechanism is now different. In the wavelengths of the radiation window, the surface is emitting directly to outside the system. In the presence of a decreasing solar input, it is therefore acting as a heat sink and absorbs longwave energy from nearby layers of air in the wavelengths outside the window. Some of the heat energy which it loses through radiation to space is brought to the surface convectively by the wind and turbulent motion still remaining in the air. The net result is that the surface is the most rapidly cooling, and later the coolest, layer in the lower layers of air. Lapse rates at the surface fall below autoconvective by mid-afternoon; isothermal conditions come near sunset; and inversion conditions soon follow. The thermal structure becomes more and more stable through the night until morning, when the cycle is repeated.

Although of no particular significance in itself, the condition of *isothermy* is sometimes used as a sort of marker, or threshold value, for examining seasonal variations of the thermal structure near the earth's surface. Isothermy is said to exist when the temperature is the same at two selected levels, one above the other, though not necessarily everywhere between. Figure 4-9 shows the relationship of the mean time of isothermy between the levels 1 and 15 meters to the times of sunrise and sunset through the year. In both hot desert and cool climates, isothermy follows sunrise by a nearly constant amount of time the year around. In a cool climate, likewise, it precedes sunset by a nearly constant amount.

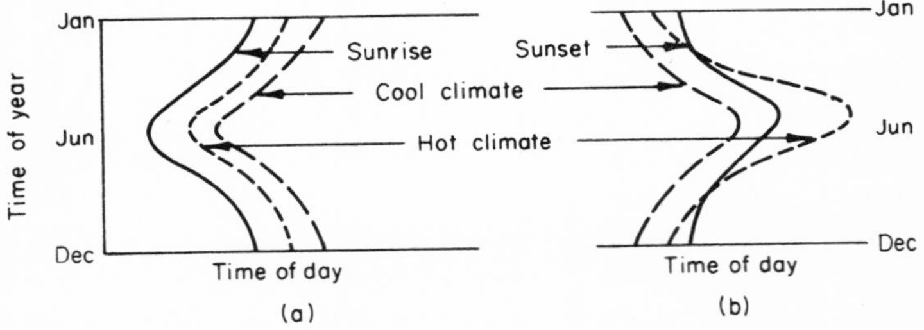

Fig. 4-9. Generalized relationship between time of isothermy in the lower atmosphere and time of year in relation to (a) sunrise and (b) sunset. (Based on Figs. 40 and 41 of Geiger,(1) 3rd ed., 1959, from data by Flower.)

In a hot dry climate, however, isothermy precedes sunset in winter and follows it by a considerable time in summer. Choosing two levels as a basis for identifying the time of isothermy, one may look at the profiles in Fig. 4-8 to see how it is that in most cases isothermy follows sunrise and precedes sunset. The greater lag of isothermy after sunrise in the cool climate is the result of longer nights in winter and of lower sun angles with attendant lower rates of heat input at all seasons. The lag of isothermy after sunset during summer in a hot dry climate may be attributed to the greater surface heating, as a result of higher sun angles and less evaporation, and thus a greater amount of cooling required for isothermy to be reached.

Mention has been made several times of the results when the soil-air system receives a greater heat input than it is capable of dissipating without significant rises in temperature of the soil surface. The nature of this overloading may be appreciated by a consideration of Fig. 4-10, which shows temperature profiles through the layer from about 1 to 40 meters before, immediately after, and several minutes after the shading of the site by a small cloud. At first, a typical midday profile existed with very high temperatures and lapse rates near the surface. The system had an "excess" heat load. As soon as the direct solar beam—and therefore the excess load—was cut off, convective and evaporative

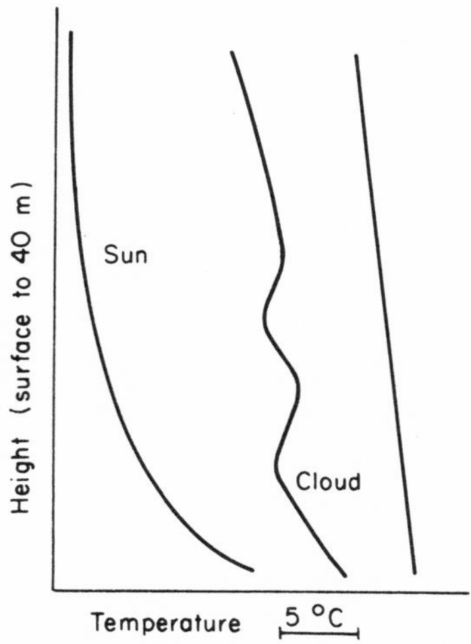

Fig. 4-10. Observed temperature profiles over grass, before, immediately following, and several minutes following arrival of a small cloud over the sun. (Based on Fig. 26 of Geiger,(1) 3rd ed., 1959.)

processes could distribute the heat load fast enough to allow equilibration. In the middle profile, the superheated air at the surface has been partially redistributed—probably by a few rising bubbles—to two or three layers above the surface. In the last profile, buoyant forces have had time enough to distribute heat equally throughout the profile. The result is a constant lapse rate in all layers: the autoconvective lapse rate.

Several discussions have shown us the patterns of temperature with height. From these profiles of temperature, one can derive information on the distribution of temperature lapse rate with height. In Fig. 4-11 are plotted the values of lapse rate against the geometric means of the heights given in Table 4-1. When these

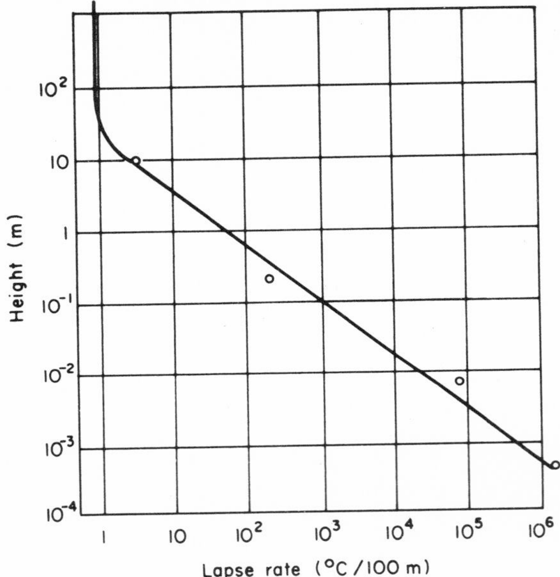

Fig. 4-11. Lapse rate of air temperature vs. geometric mean height, logarithmic coordinates, with values taken from Table 4-1.

data are plotted on double logarithmic coordinates, they form a straight line representing the equation

$$\log(\text{lapse rate}) = \log a + b \log(\text{height})$$

or

$$dT/dz = az^b. \tag{4-1}$$

This equation, empirical rather than theoretical, is often employed to present in condensed form a statement of the variation of air temperature with height near the ground. Table 4-1, and thus Fig. 4-11, represent typical midday conditions under sunny skies. The value of the empirical constant b under these conditions is usually near -1; thus the equation says that temperature is nearly a linear function of the logarithm of height. Such a relationship, as we shall see, is commonly encountered in microclimatic profiles.

Although the values of a and b in the above equation may vary with time of day and with season (i.e., with the required rate of energy dissipation), the general relationship holds remarkably well throughout a wide variety of conditions in the lowest few tens of meters of the atmosphere. Above these heights, lapse rates under conditions of good mixing almost invariably lie between the dry adiabatic and the moist adiabatic, as indicated above 50 meters or so in Fig. 4-11.

In this chapter, we have to this point described typical patterns of variation of temperature in time and space in the layers of soil and air near the earth's surface. We shall now turn attention to the explanation of these patterns in the soil, where heat transfer takes place almost exclusively by conduction. In a later chapter, similar explanations will be presented for the air layers near the ground, where heat transfer is largely convective.

Heat Conduction in Soil: Basic Considerations

The amount of heat energy (Q) which flows by conduction through a medium during time (t) is directly proportional to the cross-sectional area (A) through which the flow takes place, and to the temperature difference (dT) between the two ends of the path, and inversely proportional to the length of the path (L). That is

$$Q = k \cdot A \cdot dT \cdot t/L \tag{4-2}$$

and k is the constant of proportionality. If we concern ourselves only with a unit of cross-sectional area, $A = 1$. Further, if we adopt the notation of the calculus for small increments of time (dt), of length (dz since we shall be speaking of vertical distance from the earth's surface), and of heat flow during dt (dQ), the result is an equation whose form appears in many of the processes to be discussed later:

$$\frac{dQ}{dt} = k\frac{dT}{dz}. \tag{4-3}$$

It says the rate at which a quantity (in this case heat) flows from one level to another is proportional to the gradient of an appropriate variable (in this case temperature) representative of that quantity, with the constant of proportionality being a physical characteristic of the material in the medium through which the flow takes place. In this case, the constant has the name "coefficient of thermal conductivity," or simply thermal conductivity.

We have seen that heat flows from high to low temperature at a time rate proportional to the temperature gradient. But how can we express the heat content of a substance? To begin, we must know the *specific heat*, c, of the substance. This physical characteristic is the amount of energy required to raise the temperature of 1 gm of the substance by 1 °C, usually expressed in calories per gram per degree (cal/gm deg). If the substance of concern is not pure, but is in fact made up of two or more materials, the specific heat of the substance must be taken as a weighted mean value of the values of c of the various constituents. Let us see how that is.

We wish to work with a unit volume of a mixture, and we want a weighted mean of the specific heats of its two constituents, c_1 and c_2. It would seem logical to use as a weight the relative masses of the two substances present in the unit volume. Now the mass of a pure substance per unit of volume is called its *density* ρ and is usually expressed in grams per cubic centimeter (gm cm^{-3}). The density of a substance is, like its specific heat and its thermal conductivity, a physical characteristic of the substance.

In the mixture, we have two constituents with densities ρ_1 and ρ_2, with fractions representing their relative volumes in the mixture v_1 and v_2. In the unit of volume of the mixture under consideration, the mass of the first substance is $v_1\rho_1$ and the mass of the second is $v_2\rho_2$. Thus, the weighted mean we need is

$$c_m = \frac{(v_1\rho_1c_1) + (v_2\rho_2c_2)}{v_1\rho_1 + v_2\rho_2}. \tag{4-4}$$

Since the total mass in the unit volume of the mixture is $v_1\rho_1 + v_2\rho_2$, this is the density of the mixture, ρ_m. Multiplying both sides of Eq. (4-4) by this density gives

$$\rho_m c_m = (\rho c)_m = (v_1\rho_1c_1) + (v_2\rho_2c_2). \tag{4-5}$$

Finally, we shall note that $(\rho c)_m$ is the amount of heat energy required to raise the temperature of a unit volume of the mixture by 1°C. We see from Eq. (4-5) that this "thermal capacity" of the mixture is a weighted mean of the constituent values of ρc with relative volumes as weights. The physical units of the thermal capacity are (cal gm^{-1} deg^{-1}) (gm cm^{-3}), or (cal cm^{-3} deg^{-1}).

Table 4-2 contains values of the several physical characteristics just discussed for various materials and mixtures. Much use of these values will be made in later discussions.

We are now in a position to evaluate the heat content of a substance. A portion of the substance having a unit cross section and a depth dz has a volume dz, and a mass ρdz. In reaching a certain temperature T on the scale of absolute temperatures, the mass ρdz will have acquired a heat energy content of $\rho c \cdot dz \cdot T$. Likewise, in changing its temperature by an amount ΔT (either absolute or centigrade), it has experienced a change of heat energy content of $\rho c \cdot dz \cdot \Delta T$.

Since we have information on how heat flows in soil and how to evaluate heat stored, let us write an equation for the heat energy budget of our volume dz. The volume is depicted in Fig. 4-12.

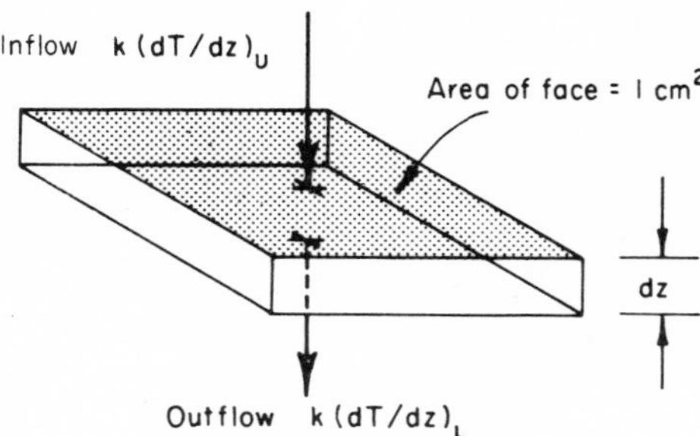

Fig. 4-12. Schematic view of a unit of soil, showing heat inflow from above and heat outflow below, leading to derivation of Eq. (4-8).

The heat flowing in through the upper face of the volume at a rate $k(dT/dz)_U$ must either be stored in the volume or flow out the lower face at a rate $k(dT/dz)_L$. Using two forms for the rate of energy input from above, this soil heat budget equation is

$$\begin{array}{ccccc} \text{Input} & \text{equals} & \text{Storage} & \text{plus} & \text{Outflow,} \\ dQ = k(dT/dz)_U & = & \rho c(\Delta T)\, dz & + & k(dT/dz)_L. \end{array} \qquad \textbf{(4-6)}$$

Table 4-2[a]

THERMAL PROPERTIES OF REPRESENTATIVE MATERIALS COMMONLY FOUND IN THE PHYSICAL ENVIRONMENT

Natural material	Thermal conductivity, k_m (cal deg^{-1} cm^{-1} sec^{-1})	Density ρ_m (gm cm^{-3})	Specific heat c_m (cal gm^{-1} deg^{-1})	Thermal capacity $(\rho c)_m$ (cal deg^{-1} cm^{-3})
Granite	0.011	2.6	0.2	0.52
Ice	0.0055	0.9	0.51	0.45
Wet sand	0.004	1.6	0.3	0.48
Wet marsh soil	0.002	0.9	0.8	0.7
Still water	0.0015	1.0	1.0	1.0
Old snow	0.0007	0.5	0.51	0.22
Dry sand	0.0004	1.4	0.2	0.3
Wood (typical)	0.00035	0.6	0.3	0.18
New snow	0.0002	0.1	0.5	0.05
Peat soil	0.00015	0.3	0.44	0.1
Still air	0.00005	0.001	0.24	0.00024

[a] Adapted from Geiger,[1] 3rd ed., 1959.

Equation (4-6) expresses the heat budget of the volume dz for a unit area of cross section and for a unit of time. The assumption is implicit that the physical characteristics ρ, c, and k do not vary with depth in the soil, and dT refers to a space difference in temperature while ΔT refers to a time difference.

Rewriting Eq. (4-6) yields

$$\Delta T = (k/\rho c)(1/dz)\{(dT/dz)_U - (dT/dz)_L\} \quad \textbf{(4-7)}$$

which in turn becomes a differential equation when the time rate of change of temperature is introduced and the dimension dz is permitted to approach zero:

$$\frac{\partial T}{\partial t} = \frac{k}{\rho c} \frac{\partial^2 T}{\partial z^2}. \quad \textbf{(4-8)}$$

The factor $(k/\rho c)$ is called the *thermal diffusivity*, and in words, the equations says that the rate of change of temperature with time at a given level in the soil is proportional to the curvature of the temperature profile at that level. In other words, other things being equal, the soil temperature will change most rapidly at those levels where the temperature profile is the least linear. Figure 4-13

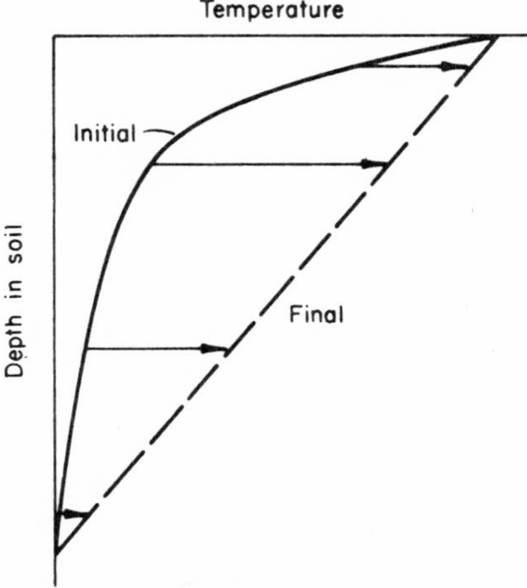

Fig. 4-13. Schematic representation (hypothetical) of the equilibration of temperature in a soil profile.

shows this relationship graphically. If we begin with a curved temperature profile, constrain the temperature at the upper and lower faces of the soil slab to remain constant with time, and allow the soil temperatures within the slab to come to equilibrium, the result will be a linear profile. During the equilibration time, the temperature will have changed most rapidly (large $\partial T/\partial t$) at the level with the greatest initial profile curvature (largest $\partial^2 T/\partial z^2$). Since at equilibrium the curvature is everywhere zero, according to Eq. (4-8), no further temperature changes will occur: equilibrium.

All of this discussion surrounding Eq. (4-8) and Fig. 4-13 pertains to ideal conditions seldom observed in the field. While our discussion may aid in understanding the basic considerations involved in soil heat flow, it takes no account of such things as the variation of thermal diffusivity with depth and time in the soil or the time variation of the rates of energy input at the soil surface. While it is very difficult to take all these variations into account, and thus to predict soil temperature changes actually observed in the field, we can go a few steps farther than we have in understanding the interactions of these various factors.

We have suggested that the thermal diffusivity varies with time and with depth in the soil. To appreciate why this is so, consider Fig. 4-14, which shows the effect of soil wetness on thermal diffusivity. In Section (1) of Fig. 4-14a, increasing soil wetness produces thin films of water around individual soil grains which when drier had thermal contact at only a few points. With only a few percent increase in soil water content, these thin films produce a large increase in the cross-sectional area of thermal contact between soil grains, as may be seen in Fig. 4-14b. Hence, with only a small increase in ρ_m and in c_m, there is a large increase in k_m: The thermal diffusivity increases.

Figure 4-14c shows the situation within the soil which pertains to Section (2) of Figure 4-14a. The soil-water-air mixture now contains a great deal more water. The films are quite thick now, and the cross-sectional area of thermal contact between grains and their films of water increases only slightly with each increment of soil water. Thus, each increment of water produces a much

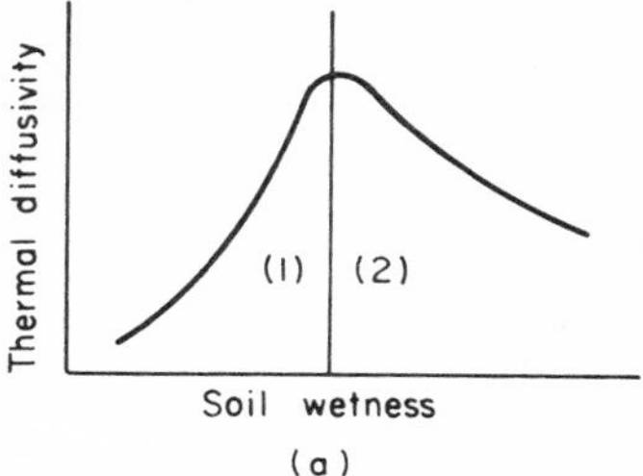

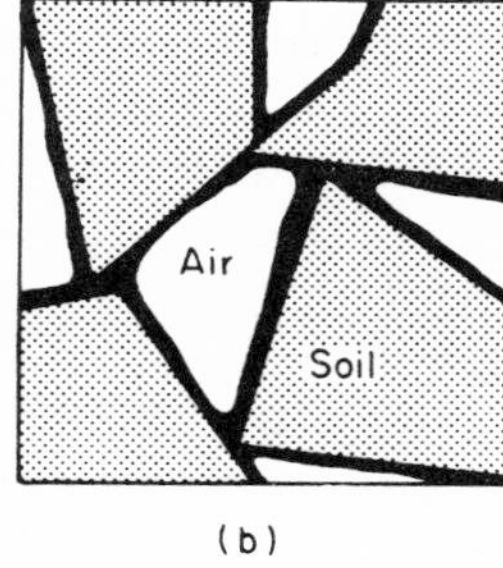

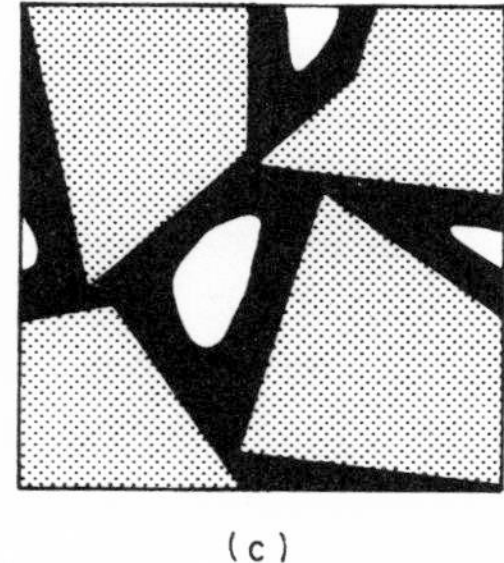

Fig. 4-14. Schematic representation of the relationship between soil wetness and soil thermal diffusivity. (a) Generalized plot of the relationship. (b) and (c) Generalized microstructure of soil-air-liquid water system.

smaller increase in k_m than in (ρc_m): The thermal diffusivity decreases.

Table 4-2 shows that the values of (ρc) for granite, water, and air—representing the three constituents of the soil mixture—are in the ratio $1 : 2 : 10^{-3}$, approximately. With the extremely small value for air, Eq. (4-5) shows we may ignore that constituent with regard to the thermal capacity of the mixture. Soil air plays an important role here, however, as a constituent with a very low thermal conductivity: an insulator. Thus, since the grain-to-grain contact area remains fairly constant with changing wetness in compacted soil, and since air is such a good insulator when motionless, changes in k_m depend on the effects of soil water films in changing the total area of thermal contact. In summary, then, as soil wetness increases from the dry end of the scale, there is first a rapid increase in k_m and only a small increase in $(\rho c)_m$. Later, in wet soil, the increases in $(\rho c)_m$ are greater than those in k_m with each increment of soil water. The resulting changes in thermal diffusivity of the mixture are shown in Fig. 4-14a.

What is thermal diffusivity, besides being a proportionality factor in Eq. (4-8)? It is a measure of the time required for a thermal impulse to travel over a given path. In terms of the soil temperature changes under discussion, it is a measure of the time for a certain kind of temperature change at the surface—for instance the rapid heating of the surface at sunrise—to be felt at a certain depth in the soil The more rapidly the soil conducts the heat (the higher is k), the sooner the impulse will reach a given depth. But the greater the amount of heat required to raise the temperatures of intermediate layers (the higher is ρc), the later the impulse at depth. Hence, the diffusivity is directly proportional to k and inversely proportional to ρc. The situation is rather like a ripple moving outward from a central point at which an impulse is introduced, and in three dimensions may be thought of as an expanding sphere starting at a point. The physical units of diffusivity are (area/time) and may be thought of as the rate at which the surface area of the sphere increases, a larger rate for a greater value of diffusivity. In the case of the soil, of course, the impulse is actually introduced on a plane (the soil surface) and the "ripples" move in plane surfaces downward into the soil mass.

Heat Conduction in Soil: The Soil Heat-Budget Equation

We are now in a position to take a few more steps in analyzing the interactions of various factors influencing heat conduction and thus temperature changes in the soil. To take these steps, we shall employ a slightly rewritten form of the soil heat-budget equation (4-6):

Input equals Storage plus Outflow,

$$dQ = \rho c(\Delta T)\,dz + k(dT/dz). \qquad \textbf{(4-9)}$$

Recall that this equation is per unit area and per unit time. Recall also that dT refers to a space difference in temperature, ΔT to a temperature difference during a unit of time. In what follows dT will refer to the temperature difference between the top and the bottom faces of the soil volume in Fig. 4-12; ΔT will refer to time changes in the mean temperature of that soil volume taken as a whole.

We shall now examine several imaginary "experiments" to see how the three parts of Eq. (4-9) change in order that the equation remains balanced. These imaginary experiments are just that: they would seldom if ever be observed under natural conditions. They are conceived and described in a manner intended to clarify the meaning of Eq. (4-9). Although semiartificial, in most respects they are near enough to reality that their principles are readily recognizable in the field.

In the cases to follow, we shall be concerned with the soil volume (Fig. 4-12) with unit horizontal area and depth dz. Its upper face may or may not lie at the soil surface. Its depth may be small or large. Although we shall speak of daytime conditions where heat is added from above and dQ is thus a positive quantity, the remarks apply equally as well to nighttime conditions with a change to negative signs where appropriate in the soil heat-budget equation. Heat in-flow from above would give dQ a positive sign, and heat outflow upward from the slab would make dQ negative. A fall in mean temperature would make ΔT, and thus the entire storage term, negative. Finally, considering greater soil depths to be equivalent to more negative values of z, (dT/dz) would have a positive value for a decrease of temperature with depth, and a negative value for an increase of temperature with depth.

The Soil-Heat Budget Equation: Case 1

To begin, assume the soil temperature profile is isothermal from the surface down through the lower face of the soil slab and beyond. Assume also a perfect insulator at the level of the bottom face of the slab. Heat is added from above, and equilibrium is allowed to take place. The results are as in Fig 4-15a: isothermal

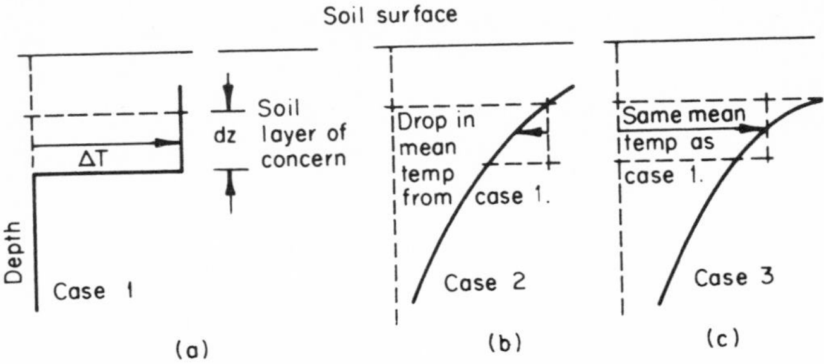

Fig. 4-15. Schematic representation of soil temperature profiles for Cases 1, 2, and 3 of the text discussion of the soil heat budget.

profiles above and below the lower face, with a temperature discontinuity at the level of the insulator. What has happened to the soil heat-budget during the addition of the heat? The following terms in the equation have *not* changed: ρc, dz, k, dT. Since heat has been added, dQ is positive. The equation has been balanced by an increase in the mean temperature of the slab: ΔT is positive. That is to say, all the added heat has gone into the storage term of the equation.

The Soil Heat-Budget Equation: Case 2

When the insulator in Case 1 is removed following equilibration, and if the temperature of the upper face of the soil slab is held constant, the results are as shown in Fig 4-15b. Since we have not added heat from above the top face of the slab, dQ is zero for the time interval between the profiles of Cases 1 and 2. Likewise ρc, dz, and k have not changed. Since the equation must balance, and since dQ equals zero, the storage term and the outflow term must equal one another. We note the mean temperature of the slab has gone down, which in terms of our equation means that ΔT is negative. The factor dT is the only one remaining. We see that dT/dz, and thus the outflow term, is positive: There is now a temperature difference between the top and the bottom faces. Thus, part of the heat held in storage at the end of Case 1 has flowed out the bottom face into the lower layers of soil.

The Soil Heat-Budget Equation: Case 3

In this case, begin as in Case 1 with an isothermal profile down through the slab of concern. This time, with the insulator no longer in the system, we add heat from above until the mean temperature of the slab is the same as it was at the end of Case 1. By adding heat until the mean temperature of our slab is the same as in Case 1, we have produced the same ΔT in both cases, as shown in Fig. 4-15c. Thus, we have stored the same amount of heat energy in the slab in both cases. As before, ρc, dz, and k have not changed. But we note that in order to store the same amount of heat energy in the slab, Case 3 required a greater addition of heat from above than when an insulator was present. By comparison with Case 1, then, Case 3 has a larger value of dQ. Since we know the storage term is the same in both cases, the conclusion must be that the increase in dQ must have been accompanied by an increase in the outflow term. The only factor remaining for consideration is dT; and we see now dT, and thus the outflow term, is larger than in Case 1 in order that the balance of the relationships be maintained. The extra heat added at the upper face has flowed out the bottom face and the added heat is thus divided between the storage and the outflow terms.

The Soil Heat-Budget Equation: Case 4

In this case, the adjustments in the soil heat budget take place within one of the terms rather than between terms as in previous cases. As in Case 3, differences are to be examined between two end results having the same starting conditions. Begin with two soil columns, identical except that one is slightly more moist than the other. Referring to the discussion of Fig. 4-14, we will say we have two soils in Section (1) where diffusivity increases with increasing soil wetness. Recall that in relatively dry soil a small increase in soil wetness will result in a much larger increase in k than in ρc. Addition of the same amount of heat from above to both columns will result in different outcomes between the two, which may be analyzed. If we assume the same division of energy between storage and outflow in the two columns, we see that between the two, the factors dQ, ρc, ΔT, and dz are essentially the same. Thus, the input and storage terms are essentially the same. But what is the result of the fact that in one case we have, by wetting the soil slightly, produced a significant increase in the size of the thermal conductivity, k? The result is that the balanced relationship between the two columns is achieved by a decrease in dT to compensate for an increase in k in the wetter soil: The wetting results in a smaller temperature gradient.

The Soil Heat-Budget Equation: Case 5

There are other ways to obtain two test soils with the same ρc and different k's. By reference to Table 4-2 and Eq. (4-5), it may be seen that a proper mixture of peat and water will have the same ρc as a wet sandy soil: 0.4. Since both peat and water have values of k less than that for wet sand, however, the value of k for the mixture will be less than that of wet sand. Though the two test soils have the same thermal capacity, the differences in their structure result in different conductivities. If we carry out the same test as in Case 4, the results will be the same: The wet peat will have a larger temperature gradient through the slab because it has a smaller value of k.

The Soil Heat-Budget Equation: Case 6

In this last case, we shall examine the results of different input rates on the same soil. Such an experiment could be conducted on adjacent plots with one shaded. The same results would be achieved if one were to have a sprinkling of white lime powder on the surface with an attendant increase in albedo. If we conduct either experiment and assume the relative division of the stored heat and outflow heat is the same in both plots, the results will be less heat stored in the shaded or limed plot and a smaller temperature gradient in that plot. In the budget equation, ρc, dz, and k are the same in both plots. The input, dQ, is smaller in the treated plot; and if the reduced input is to be shared between storage and outflow, ΔT and dT must both be smaller than in the untreated plot. The experiment with the lime powder was actually performed in India on very black soil[1] with the results shown in Fig. 4-16.

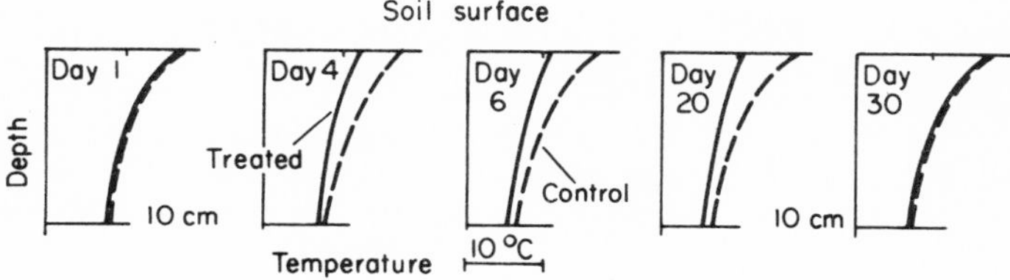

Fig. 4-16. Temperature profiles observed in an untreated soil column and in a column whose surface was sprinkled with white lime powder. Lime powder was added to the surface Day 2 and removed Day 21.

REFERENCES

1. R. Geiger, "The Climate near the Ground," p. 154 (4th ed.), Harvard Univ. Press, Cambridge, Massachusetts, 1965.

PROBLEMS

4-1. Plot the following midday temperature data* on coordinates of geometric mean height (meters) vs. temperature gradient (°C/100 meters); then answer the following (geometric mean of z_1 and z_2 is $(z_1 z_2)^{1/2}$):
(a) What height do you estimate as the lowest at which the dry adiabatic lapse rate was found at the time of observation?
(b) What is your estimate of the value of exponent b in Eq. (4-1) based on your plotted data?

Height (meters)	Mean temperature (°C)
6.4	31.23
3.2	31.86
1.6	32.44
0.8	33.43
0.4	34.41
0.2	35.25
0.1	36.24

* These and micrometeorological profile data in other problem sets are from "Exploring the Atmosphere's First Mile" (H. Lettau and B. Davidson, eds.), Pergamon Press, New York, 1957.

4-2. How does the value of the specific heat of a piece of solid granite compare with that of an equal volume of granite gravel? How do the thermal capacities of the two materials compare?

4-3. The explanations for adjustments of the temperature profile following arrival of a cloud (Fig. 4-10) and for the lag of maximum air temperature after solar noon near the ground are linked by a single phenomenon. What is the name of the phenomenon and how does it explain each of the two effects mentioned?

4-4. The amount of time by which isothermy follows sunrise (or precedes sunset) seems to be dependent upon the levels at which temperature is measured. How would you alter one or both of two measurement levels so as to observe isothermy earlier in the morning? Earlier in the afternoon?

4-5. From Table 4-2, estimate the relative volume of air in "dry sand" consisting of grains of granite.

4-6. In Table 4-2, values of thermal properties are given for dry sand (grains plus air) and for wet sand, but no indication is given of just what is meant by "wet."

(a) Calculate the ratio of volumes of dry sand and water which must have been present in the wet sand for it to exhibit the values shown in the table.

(b) If the relative volume of grains is not changed as the soil becomes wet, what were the relative volumes of water, air, and granite in the wet sand described in the table?

4-7. From the following soil temperature data (°C) estimate the two-hour mean flow and storage terms for the heat budgets of the layers 0.5 to 1.5 and 1.5 to 2.5 cm. Assume the soil's thermal conductivity is 0.001 cal cm^{-1} deg^{-1} sec^{-1} and its thermal capacity is 0.4 cal deg^{-1} cm^{-3}.

Depth (cm)	Hour				
	0400	0600	0800	1000	1200
0.5	21.04	20.81	25.68	36.05	43.47
1.5	21.68	21.23	24.40	31.88	38.11
2.5	22.11	21.56	23.77	29.45	34.86
5.5	23.05	22.37	23.05	26.02	29.90
10.0	23.01	22.60	22.37	22.92	24.44
20.0	23.12	22.86	22.58	22.46	22.70
80.0	20.22	20.26	20.27	20.28	20.22

4-8. You are keeping track of soil temperatures at several depths in a light-colored soil, and at midday you sprinkle a thin film of lampblack on the surface. Using a factor-by-factor analysis as was done in the text, deduce what changes you would expect to observe in the indicated temperatures shortly after you add the lampblack.

4-9. How much heat energy will flow by conduction in 3 hours through a column of wet marshy soil having a 10 cm^2 cross-sectional area and a 5 cm depth if the top of the column is 10°F warmer than the bottom?

4-10. If the temperature of a 10-cc pebble of granite is 20°C, what is its heat content? What volume of air at the same temperature has the same heat content? What temperature must the granite pebble have if its heat content is doubled?

Chapter 5

ENVIRONMENTAL MOISTURE

The Latent Heats of Water

OPENING CONSIDERATION of water in the environment, it is a good idea to remind ourselves of the tremendous amounts of energy involved in any change of state of water. As was noted in Chapter 2 such changes form a mode of heat transfer. When water evaporates and the vapor molecules are wafted to a different locale, the energy invested in the change of their state from liquid to gas is carried with them to be realized as heat energy at the time and place these molecules again change state.

To convert one gram of ice at 0°C to a gram of liquid water at the same temperature requires about 80 cal—four-fifths the amount of energy required to heat that same gram of liquid to the sea level boiling temperature. And then to convert this gram of liquid at room temperature to a gram of vapor requires approximately 583 cal—almost six times the amount of energy it took to heat the liquid from freezing to boiling. These energy requirements, which the changes of state in water require without changes in temperature, are the latent heats of water. The whole point in reminding the reader of these values is to underscore the fact that large amounts of energy are involved. Later we shall see that under rather ordinary circumstances frequently encountered in the microenvironment, this mode of heat transfer will completely dominate the other modes.

Vapor Pressure

A key variable in any discussion of environmental moisture is the vapor pressure of water. As the name suggests, it is the pressure exerted by the molecules of water vapor. To appreciate this physical nature of vapor pressure, consider the experiment shown in Fig. 5-1a. A sealed bell jar contains a pan of water, tightly covered,

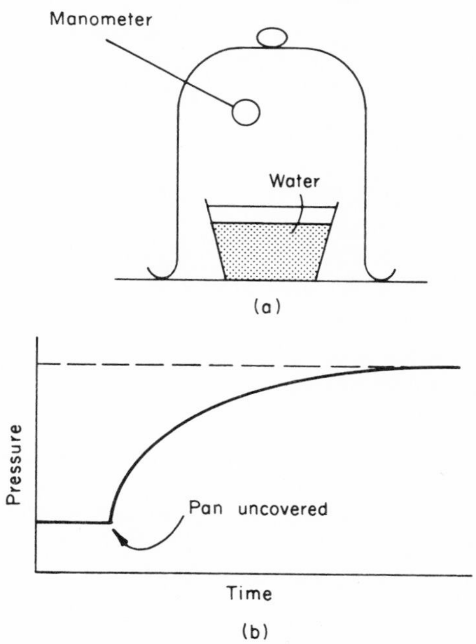

Fig. 5-1. A simple laboratory experiment showing the physical meaning of vapor pressure.

and a manometer, or pressure gauge. During the course of the experiment, the entire apparatus is kept at a constant temperature. At the outset, we note the reading on the manometer. Then we uncover the pan of water and allow water from the pan to evaporate under the bell jar. Recording the manometer readings at subsequent times will yield a plot as shown in Fig. 5-1b. The readings will increase at first quickly, and later more gradually, until no further increase is recorded. The rise in pressure is due to the force exerted on all surfaces under the bell jar by the water molecules which evaporated. When the point is reached that the number of molecules leaving the pan equals the number returning, the molecules will be exerting their *saturation vapor pressure*.

If there had been a perfect vacuum under the bell jar before the pan of water was uncovered, the reading on the manometer at equilibrium would be saturation vapor pressure, e_s, for that temperature. If perfectly dry air had been enclosed by the bell jar, the manometer would have registered the air pressure before the pan was uncovered, and the air pressure plus e_s at equilibrium. Thus, the increase in pressure during the experiment, rather than the final pressure as in the case with the vacuum, would be the value of e_s.

What if the experiment had been carried out at a different temperature? If warmer, the water vapor molecules would have greater kinetic energy and thus would exert a greater force on the manometer: The value of e_s would be higher. If cooler, e_s would be lower. The saturation vapor pressure is clearly a function of temperature. Of what else is it a function?

To bring out more clearly a point only hinted at above, consider a second experiment. Begin, as in Fig. 5-2a, with two inverted test tubes of mercury, each containing a vacuum. In tube A introduce a drop of water with a dropper. The level of the mercury will fall by an amount *X* because of the pressure of the water vapor molecules which have evaporated into the vacuum. Now, again with a dropper, introduce into tube B a bubble of dry air. The level of the mercury will fall an amount *Y*. Follow this by introducing into B a drop of water the same size as was introduced into A. The level of B will now fall below level Y by an amount *X*—the same amount as in A. If records of the mercury had been taken at various times during the fall, the records would look like those in Fig. 5-2b. While the change in level brought about by equal numbers of water molecules was the same in both tubes, the time taken to reach equilibrium was greater when the water molecules had to diffuse among air molecules already present. Thus, while evaporation rate may be a function of the number of molecules

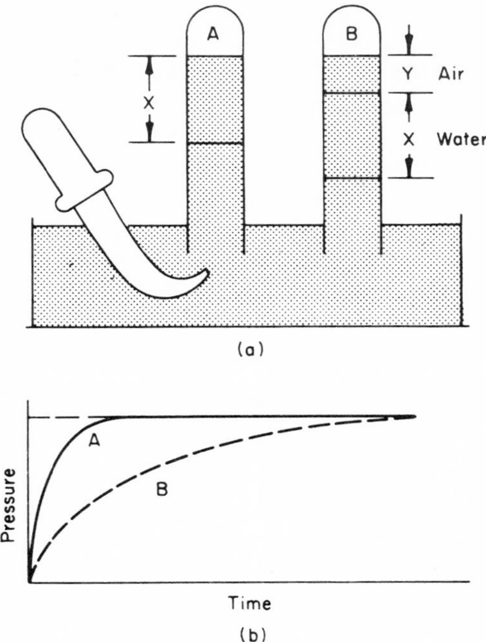

Fig. 5-2. A laboratory experiment illustrating the independence of equilibrium vapor pressures among several gases in the same environment.

of other gases present, the partial pressure of water vapor at equilibrium, and in particular e_s, is the same regardless of the partial pressures of other gases. Saturation vapor pressure is, for purposes of all discussions to follow, a function of temperature only.

It is because of the retarding effect on evaporation rate of the presence of other gases in greater concentration that, for example, food cools more slowly at lower altitudes than high in the mountains.

Measures of Atmospheric Moisture

There are various measures with which we express the amount of moisture present in a given sample of air. There are different ways of combining the factors in the ideal gas laws: temperature, volume, pressure, and mass. Referring back to the bell jar experiment in Fig. 5-1, let us assume we conducted it beginning with the bell jar filled with dry air. We then uncover the pan of water, but we close it before equilibrium is reached. We now have a sample of air containing less moisture than it is capable of holding at the temperature of the experiment. The volume of the sample we know is the volume of the bell jar. The total pressure is the sum of the partial pressure of the dry air with which we began and the partial pressure of the water vapor we allowed to evaporate. This partial pressure of water vapor is called the *ambient vapor pressure*, e. The mass of the water vapor can be calculated from the volume reduction of liquid water in the pan and the density of water at that temperature.

The first measure of atmospheric moisture present is the *relative humidity*, which is defined as the ratio of the ambient vapor pressure and the saturation vapor pressure at that temperature: e/e_s.

Another measure of atmospheric moisture is the *mixing ratio*, or the ratio of the mass of water vapor to the mass of dry air in the sample.

The *specific humidity* is the ratio of the mass of water vapor to the mass of moist air in the sample, while the *absolute humidity* is the ratio of the mass of water vapor to the volume of moist air.

Of these four measures, relative humidity is probably most often encountered in everyday life, but because its value is so dependent upon temperature it is less often used in scientific work. Mixing ratio and specific humidity are more commonly encountered in the technical literature.

Another variable employed in discussions of atmospheric moisture is the *dewpoint temperature*, T_d. If we cooled the bell jar system, containing the unsaturated air with ambient vapor pressure e, to the temperature at which the value of e was equal to the saturation vapor pressure, we would see droplets of water form on the walls of the bell jar. The relative humidity would be e_s/e_s, or 100% and the temperature at that point would be the dewpoint temperature. It is the temperature at which dew is formed when air is cooled at a constant pressure and moisture content.

Suppose in our bell jar experiment we again started with dry air, and then uncovered the pan of water and left it uncovered until saturation was reached. But this time, instead of keeping the system at a constant temperature, we allowed the air in the bell jar to cool as some of its heat was extracted to evaporate the water. The temperature reached at saturation under these conditions would be higher than the dewpoint temperature and lower than the starting temperature. It is called the *wet bulb temperature*, T_w. The difference between this temperature and the starting temperature has been related empirically to the relative humidity, and with proper tables one may obtain the relative humidity with readings from wet and dry bulb thermometers.

The Temperature–Relative Humidity–Vapor Pressure Diagram

The discussions above make it clear there are intimate connections among the many factors related to atmospheric moisture. We are in a position now to construct a diagram relating three of these factors: temperature (T), relative humidity (R), and vapor pressure (e). The *TRe* diagram will prove very useful in later discussions and is constructed as in Fig. 5-3. To begin, we have noted in the experiments performed in understanding vapor pressure that e_s is a function of T only and that as T increases, so does e_s. The relationship between them is not linear, however. It is in fact logarithmic, as shown in Fig. 5-3a. A rule of thumb for sketching this curve is that vapor pressure doubles for each temperature increase of 20°F or 11°C. It may be labeled both "e_s" and "$R = 100\%$," since both conditions refer to saturated air. But what of unsaturated air having e less than e_s and R less than 100? Figure 5-3b shows the construction of another line of constant relative humidity. By definition $R = e/e_s$ irrespective of the temperature. Thus, if we choose several temperatures, construct constant temperature lines up to the curve of e_s, and then divide these lines in the proportion e/e_s, a line connecting these proportional points will be a line of constant R. Figure 5-3c is a sketch of a complete *TRe* diagram, while Fig. 5-4 shows the same diagram with considerably more detail. An additional feature of Fig. 5-4

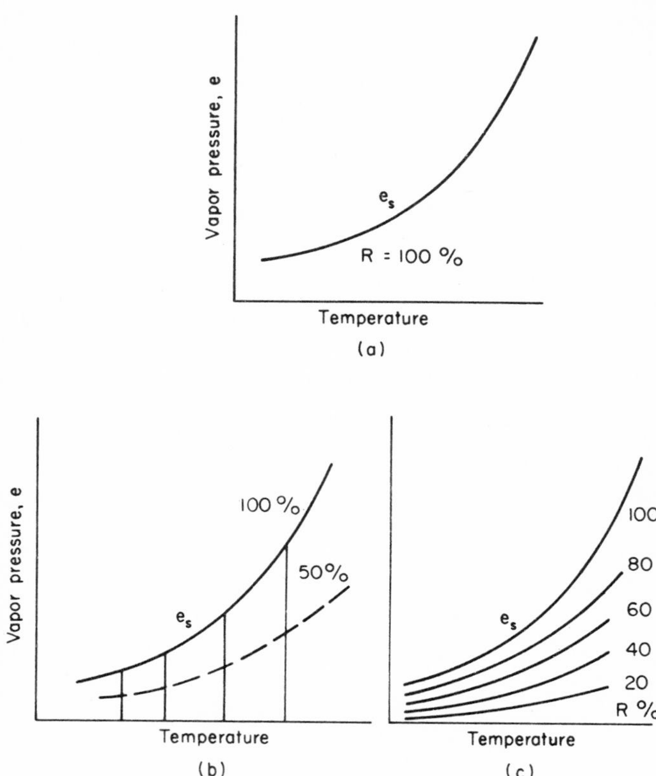

Fig. 5-3. Schematic showing steps in the construction of the temperature-relative humidity-vapor pressure diagram.

Fig. 5-4. The temperature-relative humidity-vapor pressure (*TRe*) diagram. To obtain wet bulb temperature, T_w, associated with values of T, R, and e, follow upward to the left, parallel to the diagonal straight line until it intersects the saturation curve. The set of diagonals is drawn for sea level pressure. For higher elevations, construct alternate diagonals using proportions indicated by the line marked 5000 ft.
To obtain dewpoint temperature, T_d, associated with values of T, R, and e, follow horizontally to the left parallel to lines of constant vapor pressure until intersection with the saturation curve.

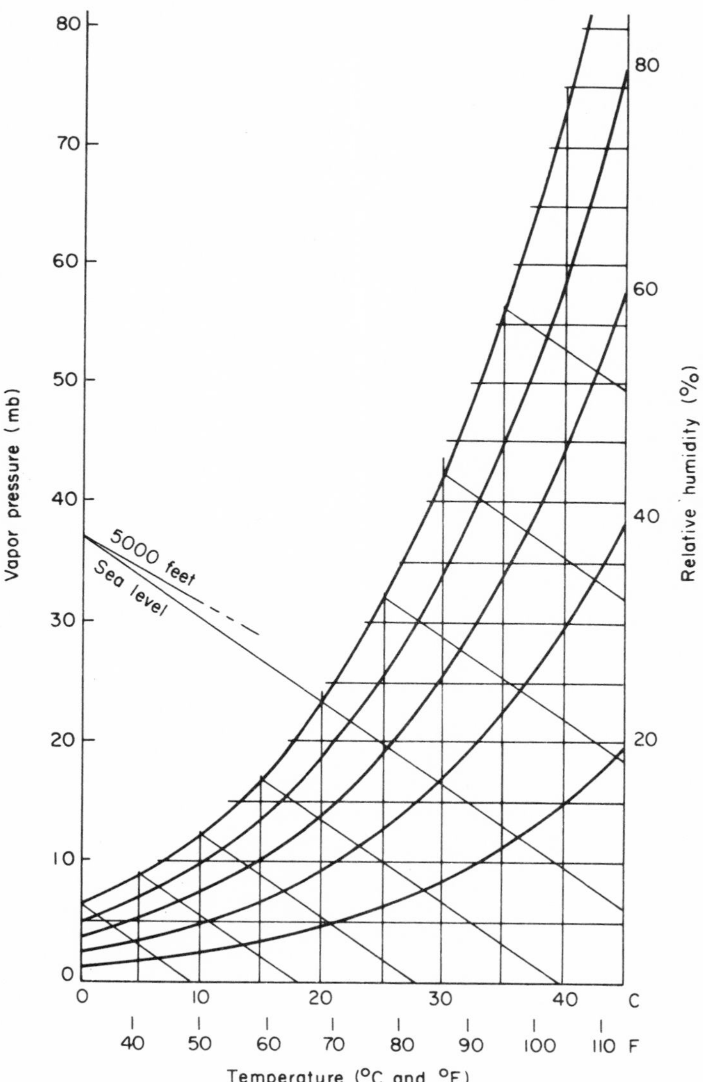

80
70
60
50
40
30
20
10
0
Vapor pressure (mb)
5000 feet
Sea level
80
60
40
20
Relative humidity (%)
0
10
20
30
40
C
40
50
60
70
80
90
100
110
F
Temperature (°C and °F)

is a set of lines for obtaining the wet bulb temperature, T_w. Knowing two of the three factors T, R, and e, one locates a point on the diagram. From there, following the diagonal straight line upward to the left will locate T_w where the diagonal meets the saturation curve. To find the value of T_d associated with the point specified by T, R, and e, one follows the horizontal line for constant e to the left until it intersects the saturation curve.

This demonstrates graphically the statement made previously that the wet bulb temperature lies between the dry bulb temperature and the dewpoint temperature.

Vapor Pressure Deficit versus Vapor Pressure Gradient

There is a rule often encountered in technical literature which says the rate of evaporation from a water source is proportional to the Vapor Pressure Deficit (VPD) of the air near the source, known also as the Saturation Deficit.

The VPD is defined as the difference between the saturation vapor pressure and the ambient vapor pressure of air ($e_s - e$). The VPD associated with temperature T_1 and ambient vapor pressure e_1 is labeled in Fig. 5-5. It will be worthwhile to examine this rule closely to see if it is correct.

Refer again to the bell jar experiment in which molecules of

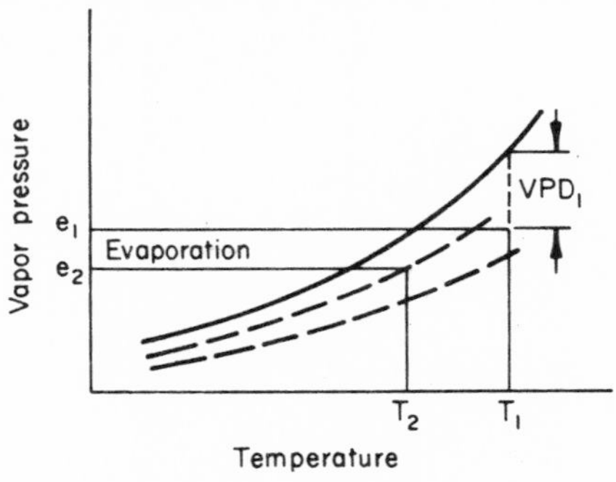

Fig. 5-5. Schematic illustration of temperature and moisture conditions at two levels near the land surface under "normal" midday conditions.

water vapor became more and more numerous in what was, at the beginning of the experiment, dry air. So long as the temperature of the apparatus was constant, the gas laws tell us, the pressure due to the vapor molecules was proportional to the number of molecules. From this we see that vapor pressure is a good measure of the density, or concentration, of water vapor molecules. We may see further how water vapor could be expected to diffuse from one place to another in air at a rate proportional to the difference in concentration of these molecules between the two places. That is to say, vapor moves from high vapor pressure to low vapor pressure. Often the term "gradient" is used in such connections: the difference in some quantity between two places divided by the distance between the two. Thus, a gradient may be doubled, for example, either by doubling the difference in concentration between two places or by halving the distance between them. Our second rule may now be expressed as follows:

> VPG Rule: Vapor moves in response to and at a rate proportional to the vapor pressure gradient, from high vapor pressure to low vapor pressure.

Returning now to the VPD in Fig. 5-5, we see that as long as the temperature is the same in all parts of a system and there is a free water source at saturation vapor pressure, the VPD rule and the VPG rule are equivalent. While such conditions, or very nearly such conditions, occur often in natural environments, there are enough important exceptions that the VPG rule is to be preferred as being more generally correct and physically more meaningful.[(1)]

Vapor Movement in the Environment

In order to become more familiar with the *TRe* diagram and with some of the natural situations for comparison of the VPG and VPD rules, begin with Fig. 5-5. Here conditions are typical of midday over turf on a sunny summer day. In this and subsequent uses of the *TRe* diagram, subscripts 1 and 2 refer to conditions very near the surface (e.g., 1 cm) and farther above the surface (e.g., 2 meters). Thus in Fig. 5-5 temperatures are higher near the surface than above it ($T_1 > T_2$), while relative humidity is higher above the surface than near it ($R_1 < R_2$). This particular combination of temperatures and relative humidities produces a gradient of ambient vapor pressure leading to evaporation ($e_1 > e_2$). While in this case we have a nonzero VPD (it can never be negative) and evaporation taking place, we note than an increase in VPD at level 1 (i.e., smaller R_1) with no change in T_1, T_2, or R_2 would produce a smaller VPG between levels 1 and 2 by virtue of a reduction in e_1. Thus, while the VPD rule holds qualitatively (VPD $= (e_{s1} - e_1) > 0$ in the presence of evaporation), it does not hold quantitatively since in this case an increase in VPD accompanies a decrease in VPG, and thus in rate of evaporation from the surface. An additional source of error in trying to apply the VPD rule comes from the absence of specification of the height at which the VPD is to be calculated. Ecologists often apply the rule to temperature and humidity data at whatever height they happen to be available. In the case of Fig. 5-5, the availability of data only at the upper level would lead to the conclusion that the VPD rule is quite satisfactory: An increase in the VPD at level 2 (decrease in R_2) results in an increase in evaporation rate [increase in $(e_1 - e_2)$].

Other examples of natural environments encountered are shown in Fig. 5-6. Figure 5-6a shows the situation which produces fog or "sea smoke" in localities such as the Grand Banks of Newfoundland or the Aleutian Islands. The same conditions are encountered over unfrozen lakes when the first subzero air mass moves southward in early winter. The underlying water is warmer than the air mass, and although the air mass is saturated ($R_1 = R_2 = 100$) evaporation is taking place. The water vapor continually condenses in the cold air, falls back into the water, and evaporates again. Here the VPD $= 0$ no matter where we calculate it, but evaporation occurs. The VPD rule has failed us.

Figure 5-6b depicts conditions found when a warm, moist air mass flows over a cold surface, as may happen in spring over a frozen lake. Here the vapor pressure and relative humidity gradients are opposite to one another and condensation is taking

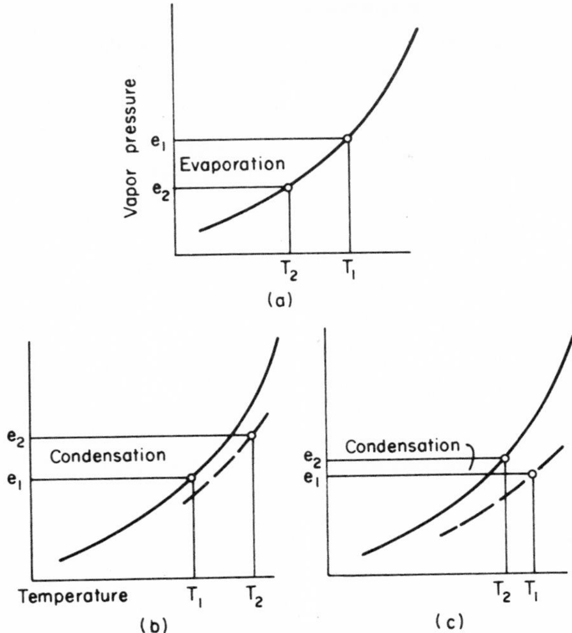

Fig. 5–6. Schematic representation of temperature and moisture stratification near the earth's surface under variously "abnormal" conditions. (a) Arctic "sea smoke." (b) Warm, moist air over a frozen lake. (c) Cool, moist air over a hot, dry surface, such as a sand dune.

place, in contradiction to the VPD rule applied at either the surface ($VPD_1 = 0$) or above it ($VPD_2 > 0$).

Finally, Fig. 5-6c shows conditions one might encounter at the surface of a sand dune or dry clayey soil on a midsummer afternoon, when cool moist air is blowing on land from the nearby ocean. The soil surface has had a chance to become very hot and dry just before the sea breeze began, and we see that so long as the high temperatures at the soil surface are maintained, water will be deposited there. In this case, the vapor pressure and relative humidity gradients are in the same direction. At either level the VPD rule is nonzero, but condensation is taking place. The VPD rule has failed again.

Moisture Lapse Rates above the Surface

It should be quite clear from the remarks just concluded that, as with air temperature, profiles of moisture in the lowest layers of the atmosphere may be quite different in different environments as well as at different times in the same environment. In addition,

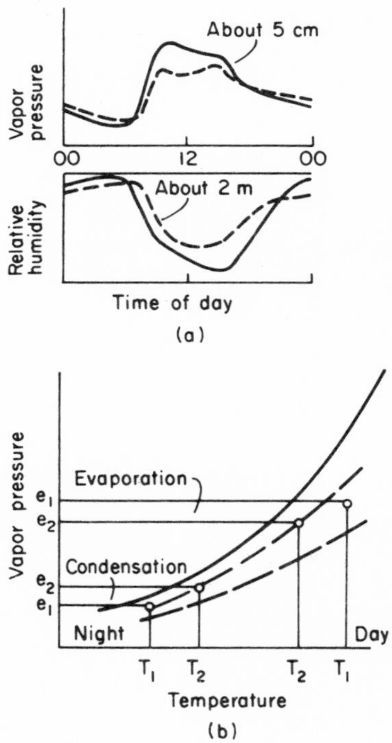

Fig. 5-7. "Normal" temperature and moisture conditions near the land surface at night and at midday. (a) Time traces, through a full diurnal period, of vapor pressure and of relative humidity at two levels. (b) Midnight and midday conditions on the *TRe* diagram.

the profile depends upon the measure of moisture being used. To underscore these remarks, consider Fig. 5-7. Vapor pressure and relative humidity at two heights (e.g., 5 cm and 2 meters) as a function of time are shown in Fig. 5-7a. Midday and midnight conditions from the time records are shown on a *TRe* diagram in Fig. 5-7b. Using terms developed with respect to temperature profiles, "lapse" will refer to a condition in which the value of the moisture variable is higher nearer the surface, and "inversion" will be used when it is lower near the surface.

In Fig. 5-7a, which depicts what might be called "normal" conditions in a midlatitude summer over turf, we see that at night there is an inversion of vapor pressure and a lapse of relative humidity. The reverse is true by day. In Fig. 5-7b, we see that qualitatively, but not quantitatively, conditions are simply reversed between day and night. The midday conditions are those found in Fig. 5-5. As with temperature, the diurnal variations of vapor pressure are greater at lower levels than at higher. But the situation is not altogether analogous between temperature and moisture. The temperature analog for Fig. 5-7a is Fig. 4-3a, but in the temperature record we find no midday minimum such as appears in the vapor pressure record for 2 meters. Why is this so?

Consider a layer intermediate between 5 cm and 2 meters—say the layer from 1 to 2 meters. In this layer, we recall from Table 4-1, midday convective mixing is fairly well developed. Thus, whatever the values of the variables recorded there, they reflect conditions both above and below the layer. The earth's surface is the midday source of both heat and water vapor as far as the lower atmosphere is concerned. In the layer under consideration, drier air is brought down from above and mixed with moister air from nearer the source at the surface. The result is an intermediate value for vapor pressure. In the morning and late afternoon, mixing is not so vigorous; drier air from above is less well represented in the layer; the vapor pressure is higher than at midday. But these same remarks about convective vigor pertain to heat, and thus temperature, as well. Why no midday minimum in the temperature record? The explanation is that heat is entering the lower atmosphere from the surface so much more rapidly than moisture, cooler air brought downward from higher layers makes very little difference in the temperature as close to the surface as 2 meters. It is quite likely a midday dip in temperature would be found at a greater height—several hundred meters perhaps—where surface-warmed air is mixed in more nearly equal proportion to air from above.

The remarks above are borne out by numerous series of observations which show that the midday minimum in vapor pressure disappears as one approaches the surface and that this minimum becomes nearly as pronounced as the nocturnal minimum farther from the surface.

Although the "normal" conditions of Fig. 5-7 are encountered in a wide variety of environments, it is not altogether unusual to find climates in which there are distinct qualitative departures from "normal." Table 5-1 summarizes conditions of vertical distribution of moisture near the ground in summer for three types of climate. While in the desert the surface is the moisture source, as shown by the fact that there is a weak lapse of vapor pressure, surface temperatures are so high and the flux of moisture from dry soil is so low that there is an inversion of relative humidity. With rapid cooling of the surface in the dry desert night, the relative humidity and vapor pressure both exhibit an inversion. In moist climates of high latitude where solar heating of the surface is never very intense, especially with dense vegetation, the flux of water vapor can keep the cool surface air nearly saturated even at midday. The result is a continuous lapse of relative humidity. In all three climates, let it be noted, the direction of actual vapor flow is upward by day and downward by night. Among these three climates only the magnitude of the vapor flux, not its direction, is altered.

Other climates could be analyzed and other kinds of remarks made about atmospheric vapor and vapor movement, but for the moment it is sufficient to understand that vapor flux and heat flux have much in common. They generally have the same directions at the same times and are mediated by the same convective processes. Both move in response to a gradient of concentration, and in both cases the variables representing concentration undergo larger diurnal variations near the surface than aloft. Finally, it is essential for proper analysis of the microenvironment to understand the physical meaning of the *TRe* diagram.

Table 5-1
A Descriptive Summary of Lapse Rates of Vapor Pressure and Relative Humidity in Various Types of Climate

	At night		Midday	
Climate type	Vapor pressure	Relative humidity	Vapor pressure	Relative humidity
"Normal"	Inversion	Lapse	Lapse	Inversion
Low latitude desert	Strong inversion	Strong inversion	Weak lapse	Weak inversion
High latitude moist	Weak inversion	Weak lapse	Strong lapse	Strong lapse

Soil Moisture: Basic Considerations

In considering the movement of water in soils, whether by liquid or by vapor flow, it is useful to describe individual types of soil by their *porosity*; the fraction of the soil volume not occupied by soil particles, or grains, themselves. It is the fractional volume available for the movement of water and air in and out of soil. According to this definition of porosity, sandy soil has a porosity much smaller than that of clay soil. But having the space available for the movement of water and air in clay does not necessarily mean movement will be free and unrestricted.

In a given volume of sand, there are a relatively few large soil grains with a comparatively few large voids, or pores, between the grains. In the same volume of clayey soil, on the other hand, there are many more soil grains; but they are all very small and have small voids scattered among them. Because the volume of typical voids in clay is less than that in sand, the size distribution of the individual voids is greatly different (Fig. 5-8).

The importance of the size of a soil void in water movement can be appreciated by considering how large voids and small voids are emptied of liquid water. A large fraction of the water held in a saturated void in sand can be removed before the thickness of the residual water film becomes very small. In clay by contrast,

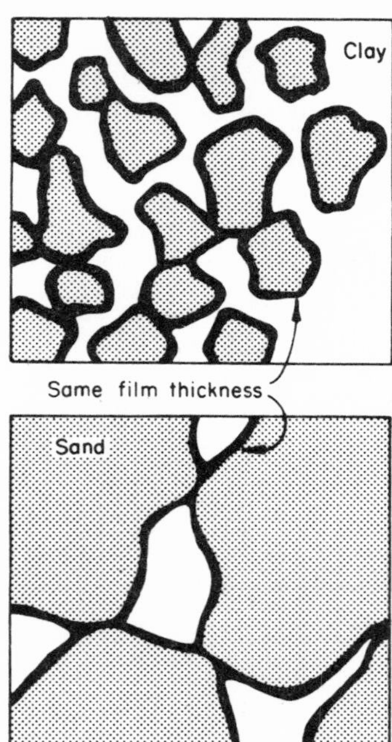

Fig. 5–8. Schematic illustration of clay and sand soil grains, with the same thin film thickness around each, showing microstructural differences in pore number and size distribution.

water has barely begun to leave when the film thickness is the same as that in our sample of sand. This relationship is also shown in Fig. 5-8. Since the resistance offered to further removal of water from a soil is inversely proportional to the film thickness, we conclude that clay, for all its porosity, can be expected to give up only a small fraction of its moisture before it commences resisting quite strongly any further withdrawals. The forces of resistance have been classified into four kinds.[2]

(1) Gravitational forces resist water movement toward the soil surface.
(2) Hydrostatic forces arise from pressures exerted by the air in soil pores as it obeys the gas laws.

(3) Osmotic forces tend to move relatively pure water from the outer edges of the film to inner portions of the film where concentrations of dissolved soil salts are high.
(4) Adhesive, or electrochemical, forces bind water molecules directly to those of soil.

The resultant of all these forces is called the *soil moisture tension*, *S*, which is in turn a measure of the energy required to remove the next unit of soil water at any given degree of dryness. The tension is expressed in units of suction, or negative pressure, usually millibars or atmospheres.

The amount of water present in soil is often expressed as the *water ratio*, *R*, or the volume of water per bulk volume of the moist soil. The meaning of this measure may be better visualized by noting it is the equivalent surface depth of all the water in a column of soil as a fraction of the soil depth. As we shall see, the value of *R* by itself tells very little about the availability of soil water for evaporation or plant use. Tension is a better unit for expressing the soil moisture's mobility.

Figure 5-9 shows the relationships between the water ratio and soil moisture tension for the two soils we have been discussing.

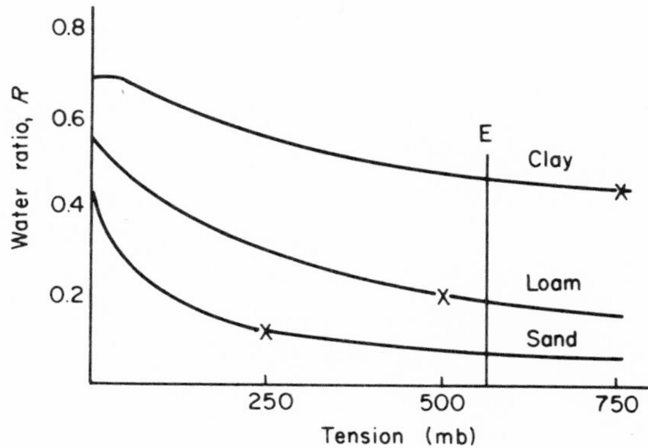

Fig. 5-9. Water ratio vs. soil moisture tension in three typical soil types. Crosses refer to the initial conditions in Fig. 5-10. The value of tension marked E refers to equilibrium conditions established in Fig. 5-10.

The greater porosity of clay appears as the much larger water ratio at low tensions when most of the voids are water-filled. The effect of the differences in pore size distribution appears in the sand's rapid reduction in R with the first small increases in S as compared with clay. At high tensions, the sand has a relatively small volume of water involved in the thin films around its few large grains, while clay still clings to most of the water it had at saturation.

Soil Moisture: Liquid or Film Flow

To appreciate the meaning of tension in problems of liquid soil moisture movement, consider the three sample soil cores placed side by side and in contact with one another in Fig. 5-10.

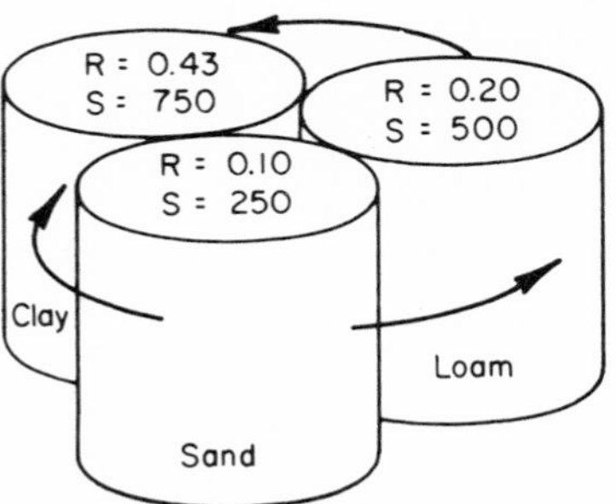

Fig. 5-10. Table-top experiment showing water movement (in the absence of gravitational force) in response to soil moisture tension. Initial conditions tabulated in soil cores match crosses in Fig. 5-9.

Each core is marked with its initial values of water ratio and tension. These initial values are also plotted in Fig. 5-9 as small crosses. With the physical arrangement we have, gravitational forces can play no role in water movement. Movement will take place according to the resultants of the other three kinds of forces in each sample, and this resultant is sometimes called the *capillary potential.* Now, what happens? Despite the fact that the highest water ratio is found in the clay, water flows from the other two soils into the clay. The explanation is that water flows in response to the gradient of tension, or "attractiveness." At equilibrium tension E in Fig. 5-9, we see that large adjustments have taken place in S but only small ones in R.

An equation relating the tension gradient to the resulting film flow takes the form:

$$dV/dt = \text{velocity} = K(dS/dn) \qquad \textbf{(5-1)}$$

where V is the volume of water passing through a unit cross section, and hence velocity when taken per unit time, dS/dn is the gradient of the tension taken in the direction n (usually, though not necessarily, vertical), while K is the hydraulic conductivity. A direct analog of the equation for thermal conductivity, Eq. (4-3), this equation states that liquid or film flow in soil responds in speed and direction to the gradient of an appropriate measure of concentration. As with its counterpart thermal conductivity, the hydraulic conductivity K has values characteristic of the various kinds of soil. In addition, as with thermal conductivity, its value may change slightly with changing conditions even within the same soil. In all the flow relationships mentioned, it is the gradient of a variable describing the energy status rather than a variable describing the relative concentration to which movement responds. Thus, it is to gradients of vapor pressure and tension rather than relative humidity or water ratio that soil moisture moves as vapor and liquid. Likewise, heat moves in response to a gradient of temperature rather than heat content.

Soil Moisture: Vapor Flow

We have seen that, at very low soil moisture contents, the liquid water is in the form of thin films surrounding soil grains. Under these conditions, there is a great deal of water vapor present in the air filling the pores of the soil. The air, of course, enters as the liquid water leaves; and since the films are free water surfaces, there is evaporation from the films into the air. We know from measurements that this soil air is very nearly saturated—in equilibrium with the water in the films—even though the water ratio is very low and the tension very high. If the air is everywhere saturated, its vapor pressure will be everywhere equal to the e_s for the temperature at that level. Since higher values of e_s will be found in regions of higher temperatures, we would expect vapor to move in response to the VPG from higher to lower temperatures. This is precisely what happens, and when liquid moisture movement is severely restricted in dry soils, vapor movement predominates. In dry soils, the high surface temperatures by day produce a flow of vapor downward into the soil—a self-mulching at times of greatest potential evaporation at the surface. At night, on the other hand, vapor moves toward the cool surface. Dew is likely to form near the surface as a result of this "distillation."

While water vapor always moves in the soil from higher to lower temperatures provided there are connecting channels of air within the soil matrix, wetter soils are less likely by virtue of higher values of ρc to develop large temperature gradients. In addition, liquid flow tends to overshadow vapor flow. Hence, vapor flow in soil becomes of major concern only as soil moisture tension becomes quite high.

Soil Moisture: Field Conditions

In an attempt to combine the ideas just presented on soil moisture and its movement, let us consider a very simplified model of what takes place under field conditions. Begin with a soil column uniformly saturated and with uniform structure and composition at all levels. Subsequent conditions will be suggested on coordinates of force (upward + and downward −) vs. depth, as in Fig. 5-11. The gravitational component is always directed downward and increases in magnitude slightly toward the surface. So long as drainage from the soil column is still taking place, the capillary potential is zero at the water table and increases positively toward the surface where evaporation is taking place (Fig. 5-11a). The algebraic sum of these two forces, and thus the resultant force on the soil water, produces a profile like the dashed line. This resultant is negative in the layers just above the water table, zero at some intermediate level, and positive above that. Thus, water

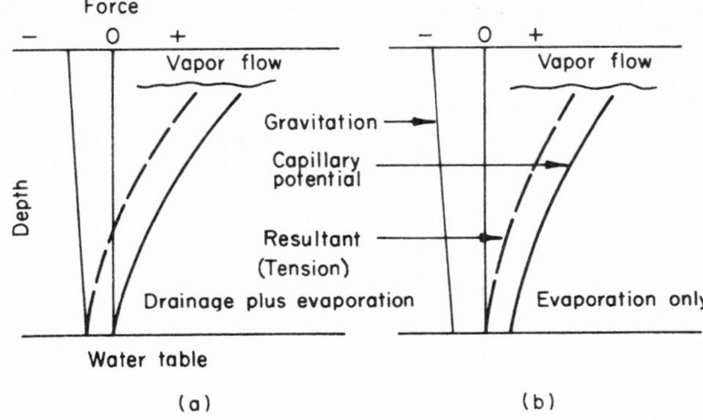

Fig. 5-11. Schematic summary of soil moisture forces and movement in a homogeneous soil column. (a) Conditions with drainage still occuring some time after initially uniform moisture distribution. (b) Conditions after drainage has ceased.

is moving downward in the lower layers and upward nearer the surface, with no net movement at some intermediate level. As time goes on, this layer of no net movement descends until it reaches the water table and drainage ceases, leaving only evaporation to operate (Fig. 5-11b).

Figure 5-11, it should be emphasized, depicts soil moisture and its associated forces only qualitatively. Relative magnitudes of the forces are not actually as shown. Likewise, Fig. 5-12 shows

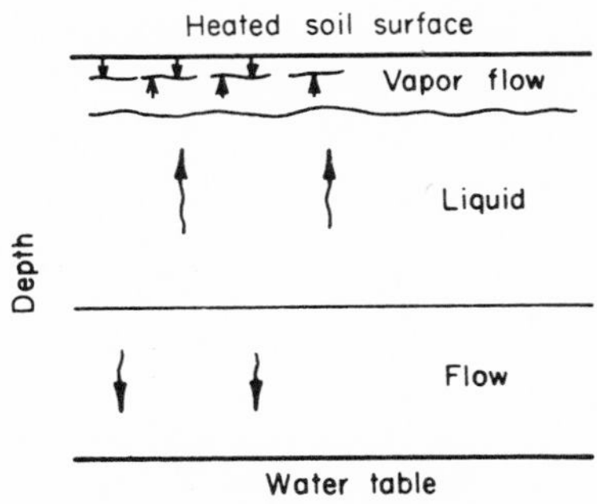

Fig. 5-12. Schematic illustration of phase and flow of soil moisture under conditions of Fig. 5-11a.

qualitatively the flow of soil moisture during the period when drainage is still taking place. We see that in the upper layers liquid flow has been replaced by vapor flow. Here the diurnal cycle of soil temperature produces a zone of zero net movement for vapor downward. At night, this zone is absent, and the gradient of e_s brings vapor to the surface, often to be condensed as dew.

Clearly, the models in Figs. 5-11 and 5-12 become enormously complicated when soil properties are different at different depths and when intermittent precipitation patterns produce several "wet fronts" moving downward into the soil. Further complication arises with the addition of plant roots to the soil water system, since roots move liquid water directly to the air above and bypass entirely the zone of vapor flow. More will be said on the subject of plant-water relations in later chapters.

REFERENCES

1. For a complete discussion of these matters, see J. Leighley, *Ecology* **18**, 180 (1937); and C. W. Thornthwaite, *Ecology* **21**, 17 (1940).
2. L. D. Baver, "Soil Physics," p. 239 (3rd ed.), Wiley, New York, 1959.

PROBLEMS

5-1. Describe an appropriate laboratory experiment which illustrates the dependence of relative humidity on temperature. You have means for measuring only temperature, pressure, and volumes of liquid water, but not relative humidity itself.

5-2. Given the following field data, calculate the difference in vapor pressure between the two levels and tell which way moisture is flowing. Now calculate what the relative humidity above the surface would be if all the other data were the same but no moisture flow were taking place.

	At surface	Above surface
Air temperature (°C)	15	25
Relative humidity (%)	90	60

5-3. For a constant temperature, is the relationship between Vapor Pressure Deficit and Wet Bulb Depression ($T - T_w$) a linear relationship?

5-4. Consider a cube 1 meter on an edge. Subdivide it into 1000 equal smaller cubes (10 on an edge) with gaps separating them so that the cubes represent soil grains and the gaps represent soil voids. Now calculate porosity of this soil model as a function of gap width.

Do the same thing for a 1 meter cube with 8 equal smaller cubes (2 on an edge).

The first soil is a model of clay; the second of sand. At what gap width do the soils have the same porosity?

5-5. Using the cube models of soil as a basis for rough calculations, develop curves of water ratio vs. tension for each of the two soils. As an expression for tension, use the reciprocal of water film thickness.

5-6. Assuming no film flow, compare rates of soil moisture movement at 0600 and at 1200 in the layers between 0.5 and 2.5 cm as described in Problem 4-7.

Chapter 6

WIND, ADVECTION, AND TURBULENT TRANSFER

IN EXAMINING WIND and its importance for biometeorology, we shall look in this chapter at the behavior of wind near the earth's surface. While this behavior is in part determined by local conditions of the other variables we have studied—radiation, temperature, and moisture—it in turn determines much of the change and distribution found in heat and moisture near the earth. The idea which we will begin to stress, then, is the *interaction* of the various components of the environment. One cannot stress too greatly the dangers inherent in "single factor reasoning" when the job at hand is to understand how the physical environment of an organism operates and how the organism fits into that environment.

Macroscale and Mesoscale Winds

We shall leave most discussion of winds occurring on a large scale to textbooks on meteorology and climatology, concerning ourselves here only briefly with some of the terms involved. Wind is atmospheric mass in motion and is caused by differences in pressure between two places. The differences in pressure are in turn caused by differences in heat content and temperature. Winds which are depicted on the scale of the ordinary weather map are called "macroscale" or "synoptic scale" winds. These are regional in nature and are responses to differences in pressure generated over distances of hundreds and thousands of miles.

Differences in the rates at which columns of air, several tens of miles apart, are heated or cooled produce pressure differences leading to "mesoscale" or "local" winds. Local winds are found near coastlines and major topographic features under conditions when temperature contrasts between land and sea and between slope and valley are substantial. Consider, for example, the land-sea breeze—one of the local wind circulations. Because of its large thermal capacity and its fluid mobility, water will not reach the high temperatures land surfaces will under conditions of strong solar heating. A column of air over the land, therefore, will heat much more rapidly on a sunny morning than a column of air over the water nearby. Through a complex chain of events, the net result of this differential heating is that winds blow from the water onshore to the land in the levels near the earth's surface—the sea breeze. The cycle is completed, under proper circumstances, by a flow from land out to sea in the air layers aloft—several thousand feet above the surface. If, during the night, the land becomes cooler than the water, a land-breeze will blow out to sea near the surface. The same sort of system often develops across the shorelines of large lakes and embayments provided their size is sufficiently great.(1)

A column of air above mountain slopes will be heated more rapidly by day than that segment of the column over a nearby valley found at the same altitude as the mountain-based column. The result is a local wind known as the "valley wind"—airflow up the slopes from the valley. The other half of the system occurs during the following night when air on the mountain slopes cools rapidly and flows into the valley—"mountain" or "drainage" winds.

Advection

Before we leave the topic of larger scale winds, it will be well to describe the nature and importance of advection which was mentioned only briefly and defined in Chapter 2 as "horizontal convection." Advection appears formally in the equation:

$$dF/dt = \partial F/\partial t + V(\partial F/\partial x) \tag{6-1}$$

(1) (2) (3)

which says the total change in a property F (term 1) during a period of time dt is the sum of the local change in that property (term 2) and the product of the space gradient of the property and the speed V with which the locality of concern is moving relative to the air in which the gradient is observed. The concept is shown diagrammatically in Fig. 6-1 on coordinates of the value of F and distance x.

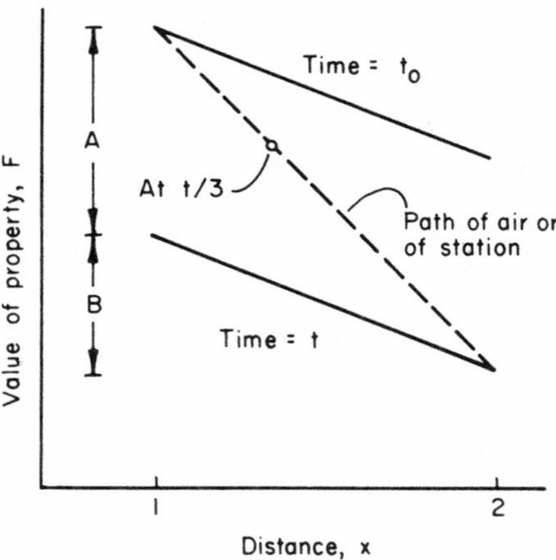

Fig. 6-1. Schematic representation of the concept of advective change in the property *F*.

On these coordinates, we first plot a line depicting the way in which F varies in space at a beginning time, t_0. Next, plot a similar line showing how F varies in space at a time t, which is later than t_0 by an amount dt. Our concern is with the way in which F varies in a locality which at time t_0 has a position 1 in space and at time t

has a position 2 on the x axis. During the time dt, our locality has followed the path of the dashed line on the diagram of Fig. 6-1, and the change in F during that time is made up of the change which has taken place at any fixed locality during dt (A in the figure) and the change due to the fact that our locality has been moving in space during dt. A is the "local change"; and B is the "advective change." Note that if there is no spatial difference in F at any given time ($\partial F/\partial x = 0$) or if the locality with which we are concerned is not moving relative to the air mass ($V = 0$), or both, the advective change must be zero. In that case, whatever change is observed must be due to local processes operating on the value of F. F may be any measurable property—temperature, moisture, dust content, ozone content, or the like. In any of these properties and many more, advective changes are often of utmost importance because they represent those changes in the record of a variable which cannot be accounted for by local processes. An example of the effect of advection and its interpretation in the temperature record at a station may be seen in the next example.

Figure 6-2 shows a temperature record from a coastal station

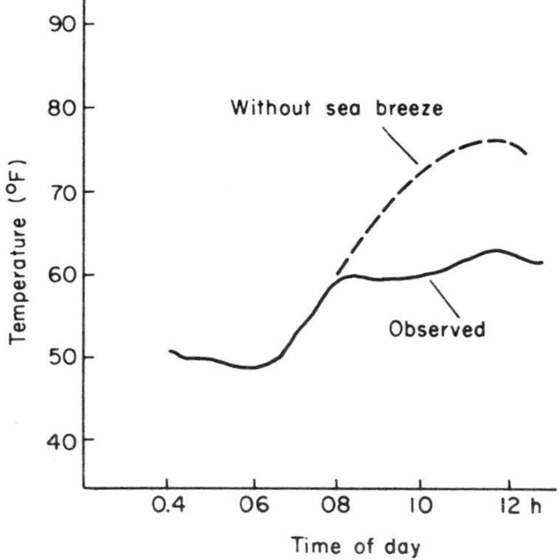

Fig. 6-2. Representative morning temperature trace for a coastal station with and without sea breeze advection.

in summer. The temperature rises from an overnight low of about 50°F at 6 A.M. to near 60°F at 8 A.M. Then, just when we would ordinarily expect an even more rapid temperature increase as the sun rises in the sky, the temperature increase ceases altogether. Has the sun gone behind a cloud? Is all the sun's energy being diverted into evaporation? These are local processes, but the explanation lies in advection borne on the sea breeze. Let's see how it looks quantitatively with the aid of Eq. (6-1).

First of all, the local change in property F (in this case, F is the temperature T) may be estimated from the rate at which the temperature had been rising just before 8 A.M.: 10° in the 2 hours from 6 to 8 A.M., or +5°/hr. In all likelihood the sea breeze at 8 A.M. had a speed of about 10 mi/hour. If the horizontal temperature gradient in the vicinity of the shoreline had been about 5° in 10 miles on the average ($\partial T/\partial x = +5°/10$ mi), we see that the advective cooling at the weather station on shore just cancelled the local change, with the result the temperature ceased rising:

$$\text{Local change} + \text{Advective change} = \text{Total change},$$

$$(+5°/\text{hour}) \qquad \begin{matrix}(-10 \text{ mi/hour}) \\ (+5°/10 \text{ mi})\end{matrix} \quad = \qquad 0.$$

In this example, the station was moving with respect to the air because the air mass was moving past the station. In another example, we can see that the advective change might arise as well when the "station"—in this case a passenger in an airplane—moves through the air.

This example may be followed by referring to Fig. 6-1, except that here we shall encounter a more realistic case in which the local changes are not the same at the two localities and, as a result, the spatial temperature gradients are not the same at times t_0 and t. Table 6-1 gives the essential information for the example. On the day of the flight, Seattle is experiencing a cool, cloudy afternoon while Los Angeles is sunny and warm.

Under cloudy skies, Seattle cools only 10° during the 2-hour flight of 1000 miles, while under clear skies Los Angeles cools 20° in the same time interval. The mean cooling rate for the two stations

Table 6-1

HYPOTHETICAL DATA FOR A SAMPLE CALCULATION OF ADVECTION

City	Temperature at 2 P.M.	Temperature at 4 P.M.
Los Angeles, Calif.	80	60
Seattle, Wash.	50	40

(local change) is $-15°/2$ hours. On boarding, the passenger is leaving a city at 80°F, and in Seattle he arrives in 40° weather—a total change for him of $-40°/2$ hours. To evaluate the advective term here, we will have to strike a mean again—this time in the temperature gradients at 2 P.M. and 4 P.M.: $-30°/1000$ mi at 2 P.M. and $-20°/1000$ mi at 4 P.M., for a mean of $-25°/1000$ mi. The equation of continuity for the example is

$$\begin{array}{ccccc} \text{Local change} & + & \text{Advective change} & = & \text{Total change} \\ (-15°/2 \text{ hours}) & & (+1000 \text{ mi}/2 \text{ hours}) & & (-40°/2 \text{ hours}). \\ & & \times(-25°/1000 \text{ mi}) & & \end{array}$$

The reader should note the algebraic signs in this equation and in the example for the sea breeze. If we assign the advance of time a positive sign, and a positive sign to travel onshore in the first case and south-to-north in the second, then we can account for the signs of terms as follows. In the example of the sea breeze, the local change was positive with an increase in temperature with an advance in time. The temperature gradient was such that for a 10 mi passage from sea onto land one would observe a 5° rise in temperature. The key to understanding why there is a negative sign for the speed is noting that the movement of the station *relative to the air* was to seaward, or in a negative direction. In the case of the airline flight, the movement of the passenger and plane *relative to the air* was from south to north, or positive. But in this second case, temperatures fell with the advance of time and temperatures were lower in the positive direction to the north.

Once again, the importance of an understanding of advection in interpretation of field records of environmental factors cannot be overemphasized. It is likely the importance will be made clearer in later chapters when evaluations of energy budgets are discussed.

Wind near the Earth's Surface: Turbulence

Many common experiences attest to the fact that wind near the earth's surface does not blow steadily and smoothly from the same direction and with the same speed.

We can feel it by facing into a gentle wind and sensing its "puffs" or "gusts" on face and body. We can see it in the nodding of grass and flower stalks, tossing of treetops, rippling of flags, and the erratic flight of a milkweed pod. To this uneven flow we give the name turbulence, and we often find it convenient to think of the motion as being made up of contributions from random movements of swirls, or "eddies," in a wide range of sizes superimposed on a basic smooth flow having constant speed and direction.

The eddies of turbulent motion are of two basic types, caused by two different processes. Air forced to move past an object protruding into the windstream will tumble and turn on itself, producing eddies having sizes and speeds related to the basic wind speed and the shape and size of the object. The result is called *mechanical turbulence*, and a detailed speed record of such a motion looks like that in Fig. 6-3a. Day and night, hot or cold, the protruding object will produce a wind record very much like this so long as the basic wind speed is the same.

The other type of turbulent motion results from the rising of superheated air bubbles and the motion of air descending to take the place of the air in the bubbles. The bubbles, and thus the eddies and gusts, usually are of a greater variety of sizes than the eddies produced in mechanical turbulence. Since these bubbles are due to heating of the earth's surface, the motion is called *thermal turbulence* and it produces a detailed wind trace like that in Fig. 6-3b. In most cases, turbulence is a mixture of the two types, but under very special circumstances the air may move without

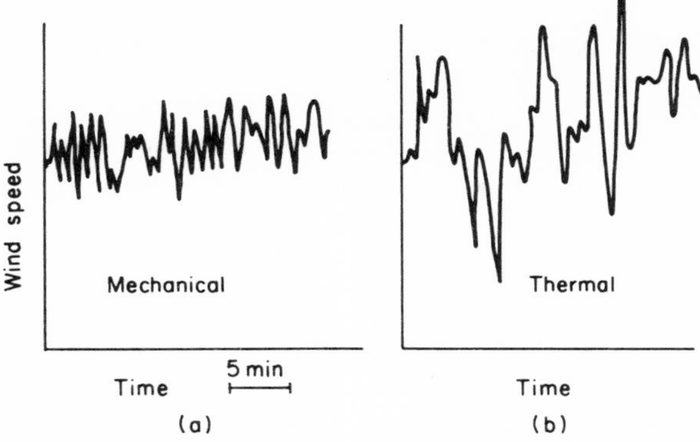

Fig. 6-3. Typical traces of a fast-response wind meter under conditions of (a) pure mechanical turbulence and (b) pure thermal turbulence.

any turbulence at all in what is called "laminar flow." Such a motion would produce a nearly straight line as its wind record trace.

Wind near the Earth's Surface: Wind Profiles

As mentioned previously, we call the curve depicting the relationship between a variable and distance from the nearby earth's surface the profile of that variable. How does wind speed vary with height near the surface?

To begin, extend the notion suggested above of packets of air moving vertically, their motion superimposed on a basic flow. For the moment, consider that the packets may be the result of either mechanical or thermal turbulence. In any case, we simplify our view of the process tremendously as follows. At some level z above the surface—several meters, let us say—the horizontal speed at any instant is the sum of two components: the basic flow, U, and the random speed being experienced just then due to an eddy, u. In our simple model we say u arises at a level z because a packet of air has moved vertically to z from a distance $\pm L$, and then has deposited at z the horizontal motion it brought with it from $(z \pm L)$. The magnitude of u is then the difference between the mean horizontal velocities at z and $(z \pm L)$. L has been called the *mixing length* and is viewed as the mean vertical path length of eddy motion, either upward or downward. In actual turbulence, the values of u exist in a range of magnitudes, but in our simple model L and u each have only one value—the average.

If the vertical gradient of mean velocity in the vicinity of z is $(\partial U/\partial z)$, then the mean speed at $(z + L)$ is $U + L(\partial U/\partial z)$. Then

(Eddy velocity at z) = (mean horizontal velocity at $z + L$) – (mean horizontal velocity at z)

or

$$u = U + L(\partial U/\partial z) - U$$

or

$$u = L(\partial U/\partial z). \tag{6-2}$$

A similar equation holds for air packets arriving at z from $(z - L)$. When Eq. (6-2) is rewritten

$$(\partial U/\partial z) = u/L, \tag{6-3}$$

we see that for a given value of u, the change in wind velocity in a given height interval is inversely proportional to L. Since large eddies and hence large values of L are associated with well developed turbulence, we expect to find under these conditions only small differences in velocity between adjacent levels.

The argument above is all very well for levels well removed from the surface, but surely we must restrict the value of L somehow near the surface. This is necessary simply because of the sheer physical barrier to vertical motion imposed by the presence of the earth's surface. A simple, direct restriction on L would be to say it is proportional to the height—i.e., the farther from the surface, the greater is the freedom to move vertically:

$$L = kz. \tag{6-4}$$

The constant k is called the von Karman constant after the famous German aerodynamicist. In various ways, its value has been determined experimentally to be about 0.4. This means, on the average, the vertical motions in turbulent eddies near the surface use about 4/10 of the room available to them for movement. Substituting Eq. (6-4) into Eq. (6-3)

$$(\partial U/\partial z) = u/kz = (u/k)z^{-1} \quad \textbf{(6-5)}$$

from which we see that the wind profile in our model has the same logarithmic form as the temperature profile of Eq. (4-1).

Although Eq. (6-5) describes the wind profile quite well in the lowest several meters of the atmosphere, for the layer from a few meters up to a few tens of meters another form of equation seems to express the profile better:

$$U_z = U_1(z/z_1)^a$$

or

$$(\log U_z) - (\log U_1) = a \log(z/z_1). \quad \textbf{(6-6)}$$

Here we have the mean speed U_z at level z expressed as a logarithmic departure from the mean speed U_1 at some reference level 1, the constant a expressing something about the intensity of the turbulence.

In Fig. 6-4a, mean wind speed observations are plotted on linear coordinates as a function of height. These speeds are mean values observed over a period of an hour or so in order to smooth out the effects of various departures such as those shown in Fig. 6-3. In Figs. 6-4b and 6-4c, the same wind data are plotted on coordinates appropriate for Eqs. (6-5) and (6-6).

A wind profile will plot as a straight line on coordinates of wind speed against log (height)—"log-linear coordinates"—under conditions described by Eq. (6-5). The curved profiles in Fig. 6-4b, then, must represent departures from these conditions. The departures arise because the model leading to Eq. (6-5) specified

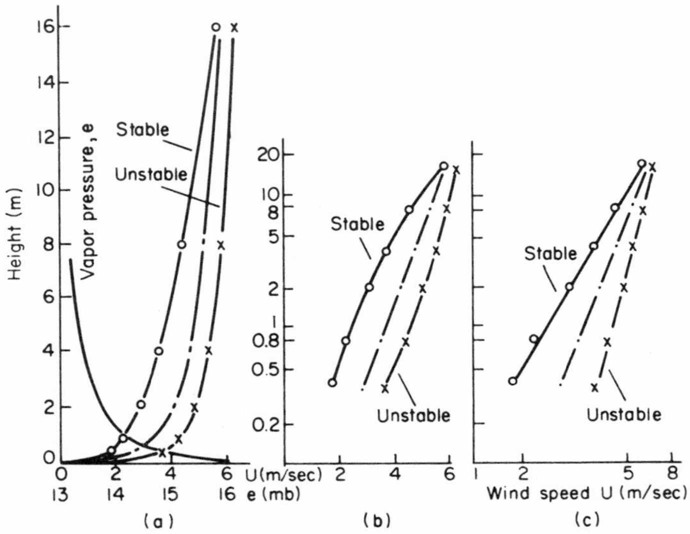

Fig. 6-4. Wind profiles in three stability classes, on three sets of coordinates. (a) Linear coordinates, (b) log-linear coordinates, and (c) log-log coordinates.

that L be equally large for downward and upward motions; that is, that buoyancy effects are not felt. Under very stable conditions —lapse rates increasingly negative—there is a tendency for sinking motion which is felt in larger mixing lengths downward than upward. The opposite assymmetry in L would be observed under conditions of instability brought about by strong surface heating. Thus, the log-linear profile of stable conditions becomes more concave downward with increasingly negative lapse rate. As the instability is increased and vertical motion is enhanced, the log-linear profile becomes more concave upward.

In Eq. (6-5), the height at which wind speed is actually zero is designated z_0. Recalling from Chapter 4 the discussion of the atmosphere's structure and behavior within a few centimeters of the earth's surface, it seems natural to think that when the surface is roughened, horizontal wind might actually disappear some small distance above the level marking the base height of the roughness elements. In terms of Eq. (6-5), the base height is $z = 0$, while the height at which horizontal wind disappears is z_0, known as the "roughness parameter."

Careful observation of a representative surface with distinct roughness elements, such as a grass turf, shows that z_0 is found above the "zero plane" where $z = 0$, and below the height of an average grass blade tip. The height where $U = 0$, then, lies somewhere within the layer of roughness elements—in this case the blades of grass. Table 6-2 gives typical values of z_0 derived from very carefully measured wind profiles above various surface materials.[2]

Table 6-2
REPRESENTATIVE VALUES OF THE ROUGHNESS PARAMETER, z_0

Surface	Value of z_0 (cm)
Fine grained sand	0.01 to 0.10
Snow	0.1 to 0.60
Mown grass	0.60 to 1.0
Short grass	1.0 to 4.0
Tall grass	4.0 to 10.0

A moment's reflection will show that windflow within a layer of roughness elements ought to depend not only upon how long, or tall, the elements are, but also on how flexible they are. Tall grass will bend in the wind. The boulders on a glacial moraine, though the same vertical dimension as grass, will not bend in the wind. This flexibility of large roughness elements is often empirically taken into account in equations of the wind profile by introduction of a so-called "zero-plane displacement," d:

$$U = (u/k)\log[(z - d)/z_0]. \qquad \textbf{(6-7)}$$

By comparison with Eq. (6-5), we see that the use of the zero-plane displacement implies a shift in the height scale by an amount d; the effect of this shift is felt largely in the few centimeters near $z = 0$, and scarcely at all at heights greater than a few centimeters.

The profiles in Fig. 6-4c show that data from all classes of stability plot as straight lines on log-log coordinates according to Eq. (6-6). Here the differences among stability classes appear as differences in slope of the lines. The logarithmic form of Eq. (6-6) shows that large differences in speed between layers imply large values of the exponent a. Since these large differences—vertical gradients—imply a suppression of interchange, or communication, between adjacent layers, we conclude that large values of a indicate large stability of stratification. It will be a good idea now to examine in more detail these ideas of interchange, or transfer, between layers of moving air.

Wind near the Earth's Surface: Turbulent Transfer

While for some purposes the presence of turbulence and the nature of wind profiles may be information of value in itself, in biometeorology these observations and processes are of interest because of the mechanisms they represent for transport of various atmospheric properties vertically. Recall for a moment the mental picture of the milkweed pod moving erratically in the turbulent air stream near the earth's surface. While the pod may move a great deal farther horizontally than vertically in a given time, it does move in the vertical a net amount under most circumstances. The net vertical motion of air parcels, as represented by the milkweed pod, is brought about by turbulent transfer—one of the major results of the presence of turbulence. Almost any measure-able property of interest in the atmosphere is moved from levels of high absolute concentration to levels of low absolute con-centration—from "sources" to "sinks." What properties? Momentum (speed), water vapor, heat, dust, pollen, ozone, carbon dioxide, carbon monoxide, and on through an almost endless list. Thus, our interest in turbulence is centered on what it does, and in wind profiles on what they tell us about turbulence.

To examine the way turbulent transfer works, let us begin by considering momentum transfer in the vertical. Momentum is the product of mass and velocity; but since, in the lowest layers of the atmosphere, the density of a given volume of air changes very little from one level to another, and because wind direction changes very little from one level to another, the momentum of a unit volume of air is almost exactly proportional to its speed. In speaking of momentum transfer, then, we are considering the way in which the property of speed is imparted by one layer of air to another by the interchange due to turbulence.

We can see intuitively that when the air at some upper level has a positive speed relative to the surface, and since at the surface the relative speed of the air must be zero, the speed is somehow reduced as we move from the upper to the lower level. We can picture this as taking place by the system of vertically moving air packets. A packet of air moving at speed $(U + u)$ moves downward from its level $(z + dz)$ to the level z. There it deposits its speed from the higher level, thereby increasing slightly the speed at level z at the expense of the speed at the upper level. Similarly, a packet at level $(z - dz)$ will move up to z and deposit the speed it brought with it from the lower level. The result will be a slight braking, or reduction in speed at z, at the expense of the lower level. We have noted that, while it is convenient to consider all these vertical trips as having the same length—namely the average—in fact the trips have a wide variety of lengths. The net result in either case is that horizontal momentum is continually transferred downward from its source in the upper air to its sink at the surface. The transfer may actually be detected at the surface. The friction between the surface and the layer of air just above it imparts a "horizontal shear stress" to the surface, which acts exactly as if there were a piece of adhesive tape stuck to a unit area of surface and the tape was being pulled in the direction of the wind. The amount of this pulling force per unit of surface area is usually given the symbol τ and is a measure of the rate at which momentum is being transferred downward in the turbulent system.

What about the turbulent transfer of a property like heat, whose source at midday is at the surface rather than aloft? The mechanism is the same as with momentum, except that each upward trip brings and deposits more heat per unit volume from the lower level than is found in the layer at the end of the trip. Similarly, downward trips reduce the heat content of the air in the layer where they make their deposits. What is true for heat at midday, of course, is true for any property having a source at the surface. On linear coordinates in Fig. 6-4a is the profile of mean wind typical for properties with source aloft. Also shown is a profile typical of those having a source at the surface. The general features of the various curves in Fig. 6-4 are practically hallmarks of conditions in the microenvironment.

Vertical Flux

The amazing result of this turbulent transfer mechanism is that in the same turbulent field of flow, in exactly the same eddies and packets and turbulence elements, and during the same periods of time, some properties of the lower atmosphere are being transferred upward and some downward. Some, like horizontal momentum, are almost always transferred downward—night and day—from upper sources to sinks at the surface. Others like dust content are almost continually transferred upward from surface sources. Heat and moisture, under most conditions, move upward by day and downward at night, the net daily *flux* through one 24-hour cycle being upward.[3]

What is the flux? It is the net, or mean, transfer of some property measured per unit mass of air and passing through a unit of horizontal area during a unit of time. Take, for example, the flux of horizontal momentum. As noted above, since momentum is (mass) × (velocity), the horizontal momentum in a unit volume of air is $\rho(U + u)$ at the level z. If we call the instantaneous vertical velocity w at the level z, then the instantaneous flux of horizontal momentum through a unit horizontal area is $\rho w(U + u)$. The time averaged value of this momentum flux is

$$\overline{\rho w(U + u)} = \overline{\rho w U} + \overline{\rho w u} = \overline{\rho w} \overline{U} + \overline{\rho w u}.$$

But since $\overline{w} = 0$, and since $\bar{\rho}$ and $\overline{U}$ are constants, the net flux is $-\overline{\rho u w}$ (downward flux is negative). Recalling that the downward flux of horizontal momentum at the surface is τ, and assuming that in the lower layers of the atmosphere τ is independent of height,

$$\text{net downward flux of horizontal momentum} = \tau = -\rho\overline{uw}. \quad \textbf{(6-8)}$$

The quantity under the averaging bar is a product of instantaneous upward and forward gust or eddy velocities. The average of these products over many instants will have a value other than zero only if there is a predominating association of upward or downward gusts with either forward or backward gusts. More specifically, τ will be positive if, on balance, most downward gusts $(-w)$ are associated with forward gusts $(+u)$, and if most upward gusts $(+w)$ are associated with backward gusts, or lulls $(-u)$.

A bit more insight into the idea of flux may be gained by a consideration of the physical units of the flux of properties, specifically horizontal momentum. As noted, the conversion of the value of a property in units per unit mass of air to units per unit volume of air is accomplished with multiplication by ρ. Calculation of flux is then accomplished by taking the time mean of the product of instantaneous vertical velocity and the instantaneous value of property per unit volume of air. For horizontal momentum, then, the units of flux are

$$\tau = -\rho\overline{uw} = (\text{gm cm}^{-3})(\text{cm sec}^{-1})(\text{cm sec}^{-1}) = (\text{gm cm}^{-1}\text{ sec}^{-2}).$$

Upon closer examination, we see that these units for τ are those of units of momentum per unit mass of air passing through a unit of horizontal area per unit time:

$$\begin{array}{lll} \multicolumn{3}{l}{\text{(Momentum/mass of air)/unit area/unit time}} \\ \text{(gm-cm/sec)} & (\text{cm}^{-2}) & (\text{sec}^{-1}) = (\text{gm cm}^{-1}\text{ sec}^{-2}). \end{array}$$

Thus, the physical meaning of flux, again, is the mean rate at which a property (in units per unit mass of air) is being transferred through a unit horizontal area.

The extension of Eq. (6-8) to the net flux of other properties is quite direct. The value of the property in units per mass of air will be analogous to $(U + u)$ for horizontal momentum. Hence

$$\text{Heat flux} = H = \rho c_p \overline{wt} \qquad (\text{cal cm}^{-2}\ \text{sec}^{-1}), \qquad \textbf{(6-9a)}$$

$$\begin{matrix}\text{Water vapor flux}\\ \text{Evaporation rate}\end{matrix} = \xi = \rho \overline{wq} \qquad (\text{gm cm}^{-2}\ \text{sec}^{-1}), \qquad \textbf{(6-9b)}$$

$$\text{Flux of property } S = \rho \overline{ws}. \qquad \textbf{(6-9c)}$$

In all these equations the instantaneous value of the property is made up of a mean value and an eddy value: $(U + u)$, $(T + t)$, $(Q + q)$, and $(S + s)$. The specific heat of air at constant pressure is c_p, so the product $c_p(T + t)$ converts temperature to heat content per unit mass of air. The vapor content per unit mass of air is the specific humidity $(Q + q)$.

We have spoken at some length about profiles and fluxes of various properties in the lower atmosphere. What is the relationship between these two concepts? The first relationship between a profile and a flux was expressed for momentum as

$$\text{Flux of momentum} = \tau = A(\partial U/\partial z) \qquad \textbf{(6-10a)}$$

which may be generalized for any property S to

$$\text{Flux of general property } S = A(\partial S/\partial z). \qquad \textbf{(6-10b)}$$

The proportionality factor A is called the Austausch, or exchange, coefficient. This coefficient is often replaced by the product ρK, where K is called the transfer coefficient. In either case, the proportionality factor is not really a constant, but instead expresses in one number the potential efficiency of the turbulent system under study for vertically transferring the property of concern. The actual values of A and K are functions of height and of the kind and magnitude of the turbulent field at work at the moment of concern. The derivative $(\partial S/\partial z)$ expresses the mean gradient of the property at the height in question. While the value of the gradient may to some extent be controlled by the process of turbulent transfer itself, it also reflects the source strength, or the amount of the property waiting to be dissipated by the turbulent field. For example, there may be a very well developed turbulent field at sunset over grassland—a large value of A—but no heat to be dissipated from the surface. Hence, near sunset isothermy ($\partial T/\partial z = 0$) is accompanied by a zero flux of heat ($H = 0$), not because of the absence of turbulence but because of the absence of a temperature gradient. The same remarks would apply to the upward flux of dust or pollen from new snow into clean air on a windy day. Plenty of turbulence and the potential for transfer, but no local source or gradient.

We have noted that A is a function of height and of the kind of turbulence present. Figure 6-5 gives information on the way A varies with height, wind speed, and time. To relate the contents of this figure to previous remarks, begin with the line representing the nighttime variation of A with U_{10} at 8 meters: the bottommost

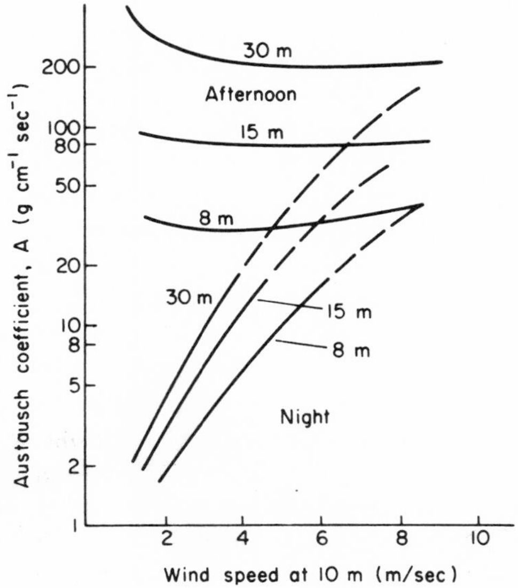

Fig. 6-5. Typical relationships among the variables of wind speed, height above the surface, and Austausch coefficient for a single value of the roughness parameter. (After Fig. 11 of Geiger,[4] from data by Frankenberger.)

curve in the figure. Clearly, A is a function of wind speed; and if we had data from other locations, we would see that A is likewise a function of the surface roughness. Under these nighttime conditions, with the coolest most dense air layers next to the surface and with no heated packets rising from the surface, we have pure mechanical turbulence with a moderate wind and nearly laminar flow with light winds. At greater heights than 8 meters, the more fully developed turbulent field represents a greater potential for vertical transfer, and thus a larger value of A for any given wind speed.

Pure thermal turbulence is represented in Fig. 6-5 by light wind conditions on a clear summer day. Here, with surface-heated bubbles growing to large sizes before rising unbroken by horizontal winds, the potential for vertical transfer is tremendous and A is very large. Then too, the proportional increase in A with height is noticeably greater than in the case of purely mechanical turbulence. These differences in A, roughly equivalent to the mixing length L, may be sensed directly in the wind records of Fig. 6-3.

Moderate winds on a sunny summer day produce a regime of mixed mechanical and thermal turbulence. The contribution made to A by thermal turbulence—the value for daytime minus the value for nighttime for a given height and wind speed—is very large for low wind speeds and decreases as winds become stronger. At speeds of 6 or 7 meters/sec (about 15 mph) at 10 meters height, the heated bubbles rise from the surface before they can become large and are further broken up by the wind until buoyancy effects on turbulent transfer are greatly reduced. The contribution of thermal elements to the Austausch coefficient is much smaller and with turbulence becoming mostly mechanical at higher wind speeds, the curves for day and night merge for a given height.

Table 6-3 is intended to gather together for side-by-side comparison the terms and concepts of turbulence, profiles, transfer, and flux developed in this chapter. Columns are arranged by environmental settings, suggesting various combinations of wind activity and heat dissipation requirements and thus vertical stability of the lower atmosphere. In any one column, the reader may make comparison of the characteristics of the various terms

Table 6-3

Summary of Parameters in Analysis of Turbulence and Turbulent Transfer

Characteristic	Night		Sunrise or sunset	Day	
	Calm	Windy		Windy	Calm
Mixing length, L (cm)	Very small	Small		Large	Very large
Austausch coefficient, A (gm cm^{-1} sec^{-1})	Very small	Small		Large	Very large
Transfer coefficient, K (cm^2 sec^{-1})	Very small	Small		Large	Very large
Kind of turbulence	Laminar flow	Pure mechanical		Mixed	Pure thermal
Thermal stratification of lower atmosphere	Stable		Neutral	Unstable	
Communication or interchange between layers	Poor	Fair	Fair	Good	Excellent
Curvature of log-linear profile (Fig. 6-4b)	Concave downward		None	Concave upward	
Exponent a (Eq. (6-6))	Large (upper limit $=1$)		0.1 (100 meter profile) 0.5 (10 meter profile)	Small (lower limit $=0$)	
Slope of log-log profile (angle with height axis in Fig. 6-4c)	Larger			Smaller	

as they exist under a given regime. Across any row, one may see how a characteristic varies under different regimes.

The units associated with the three measures of potential transfer efficiency—L, A, and K—will remain the same regardless of the property being transferred, as may be verified by careful consideration of Eqs. (6-9) and (6-10). The reader will recognize in the units of K its similarity in physical meaning to the diffusivity discussed on page 56.

As noted above, log-linear and log-log profiles are customarily employed very near the surface and a few tens of meters above the surface, respectively. The values of a given in the middle column of Table 6-3 serve further to indicate that even the value of this empirically determined dimensionless exponent depends upon the heights at which the wind measurements are made. Its qualitative behavior, however, is the same regardless of heights, as is the qualitative behavior of the other characteristics summarized in Table 6-3.

In concluding this chapter, let us examine the interconnection between profiles of one property and the vertical transfer of another—that is, the way in which movement and distribution of all the atmospheric properties are interdependent in the micro-environment. Figure 6-6 relates temperature lapse rates to wind speeds at 62 and 15 meters. The abscissa is the mean speed at the upper level, and the ordinate is the difference between the mean speeds at the upper and lower levels. If the speed (momentum) at the lower level were completely disconnected from that at the upper level, any increase in the upper level speed would result in an equal increase in the difference between levels. Under such conditions, all observations would lie along a line such as that marked 1 : 1 in Fig. 6-6. Under conditions such as those found with a temperature inversion, there is a small amount of communication between layers. Hence, data will fall on a curve wherein any increase at 62 meters results in a slightly smaller increase in the difference than that found in the absence of any communication. Higher speeds at 62 meters must result in more mechanical turbulence and thus in a curved line for conditions of "inversion." Proceeding through the range of stability classes to lapse conditions, we find that with great instability the effective communi-

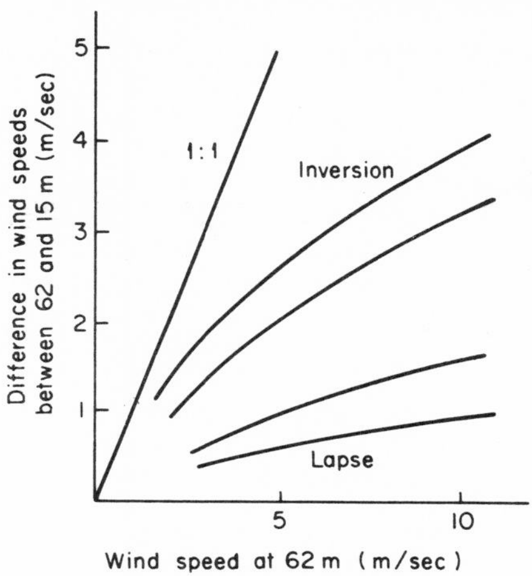

Fig. 6-6. Relationships between wind speed and wind shear (vertical difference) as a function of thermal stability. (After Fig. 59 of Geiger,(4) from data by Flower.)

cation between levels results in a transfer down to the lower levels of a large share of any increased momentum at 62 meters. Thus the wind speed under these conditions would increase proportionally at all levels and the differences would not change greatly, as in the curve marked "lapse."

REFERENCES

1. For a good discussion of local winds on a more theoretical level, see F. Defant, "Local winds," *in* "Compendium of Meteorology," p. 655, Amer. Meteorol. Soc., Boston, Massachusetts, 1951.
2. G. J. Haltiner and F. L. Martin, Dynamical and Physical Meteorology, p. 229, McGraw-Hill, New York, 1957.
3. For an excellent discussion of profiles and flux, see E. K. Webb, "Aerial Microclimate," *Meteorol. Monograph* **28**, Amer. Meteorol. Soc., Boston, Massachusetts, 1965.
4. R. Geiger, "The Climate near the Ground" (4th ed.), Harvard Univ. Press, Cambridge, Massachusetts, 1965.

PROBLEMS

6-1. You are observing nighttime temperatures in a site you suspect of being a frost pocket. Between 10 P.M. and midnight, the temperature drops from 40 to 36°F under clear skies and calm winds. Just after midnight, a breeze of about 2 mph starts to blow down valley, and the temperature drops to 30°F by 1 A.M. You attribute the marked change in cooling rate to cold air advection from up valley. Calculate an estimate of the temperature 5 mi up valley by first estimating the temperature gradient in that direction.

6-2. Given the following field data on mean wind speeds, plot profiles on linear, log-linear, and log-log coordinates. Using the equation for gradient as a function of height from Chapter 4, calculate the exponent b as a function of time from the wind profiles. Is the profile on log-linear coordinates most nearly a straight line at the same time b is most nearly equal to -1?

Height (meters)	Hourly mean wind speed (meters/sec)				
	0400	0600	0800	1000	1200
16	10.0	10.2	13.0	12.2	12.7
8	8.9	9.1	11.8	11.1	11.6
4	7.9	8.2	10.7	10.2	10.6
2	6.9	7.2	9.6	9.1	9.4
1	6.2	6.3	8.5	8.0	8.3
0.5	5.3	5.4	7.3	6.9	7.0

6-3. Following are values of horizontal shear stress, τ, at the surface, measured in conjunction with the wind profile data of the previous problem. Was the Austausch coefficient at 8 meters related to the mean wind speed at 8 meters as Fig. 6-5 indicates it should have been?

Hour	0400	0600	0800	1000	1200
Shear stress (dynes/cm^{-2})	1.7	1.8	3.4	2.6	3.3

6-4. Demonstrate that the units in the three equations (6-9) really do agree with the units required by the definition of flux given on page 102.

6-5. Develop an equation for sensible heat flux, H, as a function of the Austausch coefficient and an appropriate vertical gradient. The units for A should be the same as in the case of momentum flux and all other units should balance in your new equation.

SECTION II

Energy Budgets

Chapter 7

THE ENERGY-BUDGET CONCEPT

THE CORE of the energy-budget concept is the statement that the difference between energy inflow and energy outflow is the energy stored or utilized within the system of concern. This is the same idea encountered in the discussion of the soil heat budget in Chapter 4, except that here we will consider other systems and other kinds of energy than sensible heat. Use of such a framework for analysis of data on environment and organism, while it does not permit accounting for many important ecological interactions, it is very convenient for realistically examining the energy flows involved during selected periods of time. Such an examination is often very revealing as it sheds light on the complex ways an organism maintains itself in a given physical setting.

The Generalized Energy Budget

Work with an energy budget requires first the definition of the system to be studied. One selects the spatial boundaries and period of time to define the system, the selection usually being made so as to secure an ecologically or meteorologically meaningful system representative of a large number of localities and leading to improved understanding of some crucial biometeorological relationship.

Recalling the modes of energy transfer to be radiation, conduction-convection, and latent heat transfer, we may write the basic equation of the energy budget as

$$(S + S_o) + (L_i + L_o) + (H_i + H_o) + (E_i + E_o) + M = 0. \quad \textbf{(7-1)}$$

In this and following equations, the symbols are

- S, shortwave radiation,
- L, longwave radiation,
- H, sensible heat transfer by convection, conduction, or advection,
- E, latent heat transfer, and
- M, energy stored and utilized within the system.

The subscripts here and in what follows are i for inflow and o for outflow. In order to make the equation correct and meaningful, conventions must be established with regard to physical units and algebraic signs. Units are customarily employed representing rates of energy transfer normalized for area and time: cal/cm^2/min. The customary convention for signs is to consider all energy entering or residing in the system as positive and all energy leaving the system as negative. While clearly all terms subscripted i will be +, and all those subscripted o are −, the sign for certain groupings of terms, as we shall see, is not always so clear.

In elaborating further Eq. (7-1), we must divide each term so as to specify geometrical direction of energy transfer. Using the subscripts d, u, and h to signify downward, upward, and horizontal directions, the first term of Eq. (7-1) would be

$$S_i = (S_d + S_u + S_h)_i = (S_{di} + S_{ui} + S_{hi}). \qquad \textbf{(7-2)}$$

This expansion of all terms in Eq. (7-1) except M will result in an equation containing 25 terms instead of nine.*

In addition to subdividing terms, it is frequently advantageous to group terms, or to write equivalent terms. For example, we know from previous discussion that shortwave energy incident on an object is thereafter absorbed, reflected, or transmitted. We also know that at temperatures typical of earth, objects emit no short-

* The expanded equation is

$$\begin{aligned}(S_d + S_u + S_h)_i &+ (S_d + S_u + S_h)_o \\ &+ (L_d + L_u + L_h)_i + (L_d + L_u + L_h)_o \\ &+ (H_d + H_u + H_h)_i + (H_d + H_u + H_h)_o \\ &+ (E_d + E_u + E_h)_i + (E_d + E_u + E_h)_o + M = 0. \qquad \textbf{(7-1a)}\end{aligned}$$

wave radiation for all practical purposes. On balance, then, the only part of incident shortwave energy of concern to a system is the part which is absorbed:

$$S_i + S_o = a_s S_i = a_s(S_{di} + S_{ui} + S_{hi}), \tag{7-3}$$

where a_s is the effective absorptivity averaged over all the short wavelengths. An equivalent grouping for longwave radiation must take into account the fact that energy leaving the system is made up of the transmitted and reflected portion of the incident stream and that which is emitted by the system:

$$L_i + L_o = a_L L_i + L_{oe} \tag{7-4}$$

where the subscript e refers to energy emitted. Finally, it has been found convenient in some contexts to combine all the radiation terms into one called simply "net radiation":

$$S_i + S_o + L_i + L_o = R_n. \tag{7-5}$$

In discussions of the energy budget, it is convenient to define and use the terms "stream" and "current" in referring to individual and collective fluxes of energy. Energy *streams* are those fluxes represented by the terms of Eq. (7-1a), for instance the downward flux of incoming longwave radiation, L_{di}. Another example is the upward flux of outgoing latent heat, E_{uo}. Within each stream are *currents*, which are fluxes from specified sources or from different transfer routes. For example, within the stream L_{di} are currents from the sun, the sky, clouds, overhanging vegetation, etc. For a system containing vegetation, E_{uo} will have currents from evaporation and from transpiration. Each of these currents could be further subdivided, depending upon the requirements of the individual investigator and his problem. Transpiration, for example, could be divided into currents from upper and from lower leaves.

While closest scrutiny of the energy budget of any system would have to include that energy contained in chemical bonds of materials moving into and out from the system, most biometeorological research on energy budgets has restricted attention to radiant, sensible, and latent heat energy as outlined above. Chemical energy has been accounted for in some studies as being simply one of the components of the term M, but this metabolic energy is almost always such a small proportion of the total transfer it is disregarded as being beyond the range of sensitivity of the analysis.

Energy-Budget Systems: Type 1

The first of several types of energy-budget system we shall consider is that of a two-dimensional interface, such as a unit area of the earth's surface, land, or sea. Clearly, the objective of such an analysis is to account for all energy transferred through an interface in a particular time period. If we select a unit of area typical of some larger portion of the earth's surface, for example, there will be no horizontal streams. Since the system has no third dimension, it can have no volume and no mass. Thus

$$M = S_{\mathrm{h}} = L_{\mathrm{h}} = H_{\mathrm{h}} = E_{\mathrm{h}} = 0. \quad \mathbf{(7\text{-}6)}$$

In addition, if the hemisphere below the interface consists of soil, we could hardly expect any radiant flux from below:

$$S_{\mathrm{do}} = S_{\mathrm{ui}} = L_{\mathrm{od}} = L_{\mathrm{ui}} = 0. \quad \mathbf{(7\text{-}7)}$$

Finally, in analyses of a system in the soil-atmosphere interface, it is customary to group sensible and latent heat streams for the lower hemisphere into a single term:

$$H_{\mathrm{ui}} + H_{\mathrm{do}} + E_{\mathrm{ui}} + E_{\mathrm{do}} = B. \quad \mathbf{(7\text{-}8)}$$

The remaining terms, or streams, in Eq. (7-1a) refer to those for the upper hemisphere:

$$S_{\mathrm{di}} + S_{\mathrm{uo}} + L_{\mathrm{di}} + L_{\mathrm{uo}} = a_{\mathrm{s}} S_{\mathrm{di}} + a_{\mathrm{L}} L_{\mathrm{di}} + L_{\mathrm{uoe}} = R_{\mathrm{n}}, \quad \mathbf{(7\text{-}9)}$$

$$H_{\mathrm{di}} + H_{\mathrm{uo}} = H; \quad \mathbf{(7\text{-}10)}$$

$$E_{\mathrm{di}} + E_{\mathrm{uo}} = E, \quad \mathbf{(7\text{-}11)}$$

so that

$$R_{\mathrm{n}} + H + E + B = 0, \quad \mathbf{(7\text{-}12)}$$

which is an equation often encountered in the energy-budget literature for systems of Type 1. The factor E in energy terms, of course, is actually the product of the latent heat of vaporization (or of sublimation) of water and the mass of water evaporated (or which sublimes).

If the lower hemisphere in Type 1 is water rather than soil, the terms in Eq. (7-7) could no longer be considered negligible, though the longwave terms would be nearly so. For reasons of analytical consistency it would seem appropriate to include these streams with the B term. For reasons of instrumentation and measurement, however, they may be better handled by including them with the net radiation, R_n.

Energy-Budget Systems: Type 2

Two-dimensional systems which are not interfaces may be represented by a thin broadleaf. While such a leaf, of course, does actually have three dimensions and thus volume and mass, analyses found in the literature have chosen to ignore the mass and treat the leaf as a Type 2. If the leaf is assumed horizontal, Eq. (7-6) applies. With a different lower hemisphere from that in Type 1, however, Eqs. (7-7) and (7-8) are not appropriate, and a regrouping of the remaining radiation terms of Eq. (7-1a) is accomplished as follows:

Upper hemisphere

$$S_{di} + S_{uo} = a_s S_{di} \tag{7-13}$$

$$L_{di} + L_{uo} = a_L L_{di} + L_{uoe} \tag{7-14}$$

Lower hemisphere.

$$S_{ui} + S_{do} = a_s S_{ui} \tag{7-15}$$

$$L_{ui} + L_{do} = a_L L_{ui} + L_{doe} \tag{7-16}$$

In order for the investigator to examine the various modes of heat dissipation from a leaf, the L terms are customarily separated into those involved in emission from the leaf and those not so involved:

$$S_i + S_o + L_i + L_o = a_s(S_{di} + S_{ui}) + a_L(L_{di} + L_{ui}) + (L_u + L_d)_{oe}. \tag{7-17}$$

Rewriting the emission terms by means of the Kirchhoff and Stephan–Boltzmann laws, and grouping the sensible and latent heat terms by hemisphere, produces the following working equation for a Type 2 system:

$$a_s(S_{di} + S_{ui}) + a_L(L_{di} + L_{ui}) + 2\varepsilon\sigma T_1^{\,4} + (H_u + H_d) + (E_u + E_d) = 0 \quad \textbf{(7-18)}$$

where T_1 is the leaf temperature, σ is the Stephan–Boltzmann constant, and by Kirchhoff's law, $\varepsilon = a_L$. In this working equation, the first two terms represent the radiant heat load on the system, and the last three terms represent the three modes of heat dissipation: radiation, convection–conduction, and latent heat.

Energy-Budget Systems: Type 3

This type of system may be described as a three-dimensional interface. With upper and lower hemispheres consisting of different media, the system is typified by a section of plant canopy, with upper and lower faces being a unit area and depth being that of the canopy. If the system is part of a continuous and homogeneous canopy of large horizontal extent, the *h* terms in Eq. (7-1a) may be ignored. If the system is near or includes a stand border, these terms may be very important and advection must be accounted for. With three dimensions, the storage and utilization term must be included, and may be subdivided into biological, sensible and latent heat in organisms (mostly plants, but also animals if they are present and important) and canopy air:

$$M_b + (M_H + M_E)_p + (M_H + M_E)_a = M \quad \textbf{(7-19)}$$

where

b refers to metabolic and photosynthetic energy,
H and *E* refer to sensible and latent heat respectively, and
p and a refer to organisms and canopy air respectively.

Since vapor in organisms is such a small fraction of the total water in the organisms, we may ignore M_{Ep}. Considering the upper and lower hemispheres to be those above the canopy and below the soil surface, we may handle the lower hemisphere as in Eqs. (7-7) and (7-8) and combine the ideas of Eqs. (7-9) through (7-12) and Eq. (7-18) to get a working equation for Type 3[1]

$$\begin{aligned} &a_s S_{di} + a_L L_{di} + \varepsilon\sigma T_c^4 + H + E + B && \text{(vertical fluxes)} \\ &\quad + (H_h + E_h) && \text{(advection)} \\ &\quad + M_b + M_{Hp} + (M_H + M_E)_a = 0 && \text{(stored and utilized)} \end{aligned} \quad \textbf{(7-20)}$$

where T_c is the effective radiating temperature of the canopy system, and the horizontal streams of radiation are assumed negligible. Tracing the fate of energy within the canopy would require such evaluations as the albedo of the soil surface as compared with that of plant materials, the fraction of the horizontal area occupied by leaves at any depth in the canopy, the evaporative as against the transpirational currents of latent heat, etc. This intracanopy analysis is simply bypassed by Eq. (7-20), but for some problems it may be essential to approach it directly.

Equation (7-20) and the considerations behind it are quite appropriate for the study of a Type 3 system consisting of a stand of vegetation in shallow water rather than air. The temperature for upward emission would be that of the water surface, and the storage and utilization factors would appear more suitable if $(M_H + M_E)_a$ were changed to simply M_{Hw} for the change of canopy medium. Otherwise, Eq. (7-20) is suitable as it stands.

Energy-Budget Systems: Type 4

Under this type we shall consider three-dimensional systems which are not in an interface, which is to say they do not clearly form a zone between two dissimilar hemispheres. Examples of Type 4 systems are an individual building and an individual organism, such as a man. Because these systems are not units taken to represent larger systems having great horizontal extent and uniformity, it is more appropriate that the energy budgets should not be normalized to unit area, but that they be taken with respect to the entire system. For the same reason, elaboration by geometrical direction is no longer appropriate: vertical and horizontal fluxes are combined.

With physical units for the S, L, H, and E terms in (cal cm^{-2} min^{-1}) as in Types 1, 2, and 3, Eq. (7-1) must be modified by proper area multiplication to be appropriate for Type 4:

$$(S_i + S_o)A_s + (L_i + L_o)A_L + (H_i + H_o)A_H + (E_i + E_o)A_E + M = 0 \quad \textbf{(7-21)}$$

where the A's are the appropriate areas involved in the various transfer processes.

If the system is a heated building, the conversion of the fuel to heat within the building may be included as a term within the M term not encountered in the previous systems. The same treatment may be due to metabolic energy in the case of a system consisting of a living animal. In the case of an animal, as in the Type 3 case of a canopy under water, the energy involved in system air will be negligible.

Energy-Budget Systems: Concluding Remarks

In this chapter we have set forth the general concept of the energy budget, and we have elaborated it to four progressively more complex types of system. This treatment has been adopted to give a feeling for the scope and applicability of the concept rather than as a definitive set of system types. A few moments of thought will show that there may be considerable overlap among types, and thus a requirement for the investigator to adapt the notions and notations of this chapter to other problems. For example, while a leaf may be considered as a Type 2 system in itself, it also forms part of a canopy the energy budget of which is the integral of the budgets of all the leaves and other plant parts. Conversely, one may treat a canopy as a Type 1 system by analyzing the budget of a unit area of the horizontal plane just above the canopy. Then too, one may treat a collection of buildings or of organisms spread out horizontally as a kind of canopy, thus turning a Type 4 system into a Type 3 system.

Finally, an investigator approaching an energy-budget problem for the first time will quickly realize the specific treatment given his data will depend in large measure on the instruments he has available: how they work and what adaptations he can make of them. The rewriting of $(L_u + L_d)_{oe}$ as $2\varepsilon\sigma T_1{}^4$ in Eq. (7-18) is an example of this, since it is a great deal easier to find and use an instrument for measuring leaf temperature than for measuring longwave radiant energy flux from a leaf directly. Besides, the leaf temperature, as we shall see in the next chapter, is also useful in the evaluation of both the H's and the E's in Eq. (7-18). In addition, the reader will notice that the "back radiation" term expressing the emission from a system, i.e., some multiple of $\varepsilon\sigma T^4$, is included within the R_n term of the Type 1 system and is treated separately as a dissipation term in the working equations of other types. This is just one of many confusing results of the historical development of both instruments and the energy-budget literature.

These last remarks will serve to introduce the next chapter, where we shall examine the particular techniques, results, and interpretations which the literature provides for several types of system.

REFERENCES

1. This is essentially the equation for a canopy discussed by V. Suomi and C. B. Tanner, *Trans. Amer. Geophys. Union* **39**, 298–304 (1958).

PROBLEMS

7-1. Write equations for S_o, L_i, and L_o similar to Eq. (7-2) for S_i. From these, derive Eq. (7-3) and (7-4). Now satisfy yourself that Eq. (7-5) follows logically.

7-2. Show how each of the terms of Eq. (7-1a) appears in Eq. (7-20) or has been set equal to zero.

Chapter 8

ENERGY BUDGETS OF PARTICULAR SYSTEMS

Before examining the factors which determine how a system's heat-energy load will be dissipated, let us examine those which determine that heat load. Table 8-1 is arranged to show streams, currents, and the variables determining the flux magnitudes for energy incident on a Type 2 system. The lists are not exhaustive and modifications would be necessary for each of the other types of system, but the table as it stands serves to summarize many of the ideas previously presented and to reemphasize the complexity of the incident heat load. The role of precipitation in an energy budget has not been discussed up to now. Suffice it to say, the water, ice, or snow brings sensible heat proportional to its mass and temperature; and that it represents a negative source (i.e., a sink) for latent heat. Transpiration and its controlling factors will be taken up in the next chapter.

Importance of System Surfaces and their Temperatures

As noted previously, the primary response of a system to a heat overload is a rise in surface temperature, until a new equilibrium is reached at which heat dissipation processes can keep up with the heat load. Likewise, surface temperatures fall if the heat load is less than can be dissipated with present surface temperature.

Rising surface temperature increases the radiant dissipation in accordance with the expression $\varepsilon\sigma T^4$, the sensible heat dissipation

Table 8-1
FACTORS THAT DETERMINE ENERGY INCIDENT ON A TYPE 2 SYSTEM

Stream	Current or source	Depends upon ...
S_{di} (and S_{hi})	Direct solar	Sun angle, atmospheric turbidity, elevation of site
	Sky scatter	
	Cloud reflection	Cloud type and amount
	Other reflections from environment	Albedos of nearby surfaces
S_{ui}	Reflection from underlying surfaces	Sun angle, albedo of underlying surface
L_{di} (and L_{hi})	Water vapor (bands)	Amount and temperature of vapor
	Other gases (bands)	Amount and temperature of gases
	Clouds (all wavelength)	Cloud amount, temperature
	Vegetation (bands)	Vegetation density, height, temperature
	Nearby slopes (bands)	Nature and temperature of slope materials
		Enclosure angles determining the field of view connecting system and each source
L_{ui}	Surface objects	Nature and temperature of surface materials
H_{di}, H_{ui}	Convection	Temperature gradients and exchange coefficients, air density
E_{di}	Vapor flux	Specific humidity gradient, exchange coefficient
	Precipitation	Precipitation type, amount, temperature
E_{ui}	Evaporation	Soil type, wetness, temperature
	Transpiration	Plant temperatures, water status
		Specific humidity gradient, exchange coefficient

by increasing the temperature difference between surface and air (and thus the temperature gradient near the surface), and the latent heat dissipation by increasing the vapor pressure of a moist surface (and thus the vapor pressure gradient near the surface). Thus, all three modes of dissipation are temperature-sensitive, but in different proportions: the radiant according to the fourth power of temperature, the sensible linearly according to the temperature rise, and the latent exponentially according to the shape of the saturation vapor pressure curve of the *TRe* diagram.

While temperature rises will produce dissipation of a heat overload on any surface, the amount of the rise required to reach an equilibrium is determined by various physical characteristics of the surface materials; such as albedo (color), longwave absorptivity (and thus emissivity), roughness (surface area), and so on. The albedo may easily be seen to affect the magnitude of the heat load, while the emissivity determines how much additional energy may be dissipated by each increment of temperature rise. The surface roughness affects the albedo, the sensible heat transfer (by affecting the turbulent exchange coefficient), and the latent heat transfer (by enlarging the surface vapor-exchange area).

Energy-Budget Type 1: Estimation of Components

In this type of system, the energy budget is determined by evaluation of the terms of the equation

$$R_n + H + E + B = 0. \tag{8-1}$$

The radiation term here expresses the heat load; and the sensible flux, latent flux, and soil-heat flux represent the modes of dissipation. Let us first examine several commonly employed methods for evaluating each of these four terms.

Usually it is not necessary to evaluate the various streams making up the net radiation, since single instruments calibrated to indicate the net radiation directly are readily available. The variety of radiation instruments commonly used, however, makes it possible to obtain values for the four vertical radiation streams if necessary. Table 8-2 names four of the types of instrument often used, together with the combination of streams each measures. Finally, the algebraic sums of instrument output are given for obtaining each separate stream.

Table 8-2

TYPES OF RADIATION INSTRUMENT AND THE FLUXES EACH MEASURES

Type of instrument	Flux measured
All-wave hemispherical	$F(A) = S_{di} + L_i$
All-wave net	$F(B) = (S_{di} + L_{di}) - (S_{uo} + L_{uo})$
Solarimeter	$F(C) = S_{di}$
Thermal radiometer	$F(D) = L_{uo}$

To obtain individual streams:
$S_{di} = F(C)$; $L_{di} = F(A) - F(C)$;
$L_{uo} = F(D)$; $S_{uo} = F(A) - F(B) - F(D)$

If one places temperature sensors at various depths in the soil, and evaluates the heat storage in each soil slab bounded by a pair of adjacent sensors according to

$$B_\mathrm{i} = \{\rho_\mathrm{m} c_\mathrm{m}(\Delta T)\, dz\}_\mathrm{i}/(\Delta t), \tag{8-2}$$

the soil heat flux, B, will be the sum of the B_i's for all the slabs down to the depth where no temperature change takes place during the observation period. We see, therefore, that R_n and B may be evaluated fairly directly without undue complication.

Unlike the soil heat flux in a solid, stationary medium, the fluxes in the atmosphere take place primarily by convection in a moving medium. Since under these conditions there is mass movement between adjacent slabs, or layers, of air, the problem of obtaining reasonably accurate and reliable values for H and E is more complex. There are three fundamentally different approaches to these evaluations.[1]

First, if a typical volume of the soil below the system can be isolated from its surroundings so its weight may be obtained as a function of time, weight changes may be appropriately converted to evaporation rates and thus to latent heat flux. Such an isolated soil mass is called a "lysimeter," and its construction and maintenance may be quite simple or quite complex. With R_n, B, and E evaluated, H is obtained as an algebraic residual of the other three.

Second, very accurate and fast response instruments may be installed at some single level above the surface so as to produce measurements of vertical wind speed, temperature, and specific humidity as functions of time. From these data, the appropriate fluxes may be obtained from the equations

$$H = \rho c_p \overline{wt} \qquad \text{and} \qquad E = L_v \xi = \rho L_v \overline{wq} \tag{8-3}$$

where L_v is the latent heat of vaporization for water.

To obtain mean values of the products, of course, requires use of many instantaneous values of departures from means. Clearly, the required instrumentation is almost prohibitively complex and delicate for most studies; and the data reduction and computation tasks are almost as demanding.

Finally, a class of methods for evaluating H and E employs in various ways measurements of temperature, moisture content, and wind speed taken at several levels in the air above the system. One of these methods employs the so-called Bowen ratio, β, which is the ratio H/E. In the form relating fluxes to profiles, the ratio is

$$\beta = \frac{c_p K(\partial T/\partial z)}{L_v K(\partial q/\partial z)} = \frac{c_p}{L_v} \cdot \frac{(T_1 - T_2)}{(Q_1 - Q_2)} \tag{8-4}$$

where the temperature difference and specific humidity difference are those observed in the mean between levels of observation 1 and 2. This method, therefore, assumes that the energy to be dissipated by H and E is partitioned between these two modes according to the value of β.* With values for R_n and B obtained as above, the evaluation of the energy budget is complete.

Another method employing measurements at two levels is due to Thornthwaite and Holzman[2] and involves evaluation of E by the equation

$$\xi = \rho k^2 \frac{(U_1 - U_2)(Q_1 - Q_2)}{[\ln(z_2/z_1)]^2} = \frac{E}{L_v} \tag{8-5}$$

where U's and the Q's are time mean values of wind speed and specific humidity at heights z_1 and z_2.†

Once E has been evaluated by this relationship, which is reliable only for conditions of nearly neutral stability, the value of H is the residual of R_n, B, and E. Besides the requirement for neutral stability, another requirement from the derivation is that z_1 and z_2 be well above the surface—the lower level at least, say, 2 meters—where isotropic turbulence may be found. Unless extremely accurate humidity sensors are available to measure what are usually very small gradients of Q, the interval between z_1 and z_2 should be of the order of 5 meters.

Table 8-3 gives a set of observations for two levels of instruments and pertinent values for other factors in Eqs. (8-4) and (8-5). From these, illustrative calculations of the two components, H and E, of the Type 1 energy budget may be made.

Using these values and observations yields

$$\beta = \frac{c_p(T_1 - T_2)}{L_v(Q_1 - Q_2)} = \frac{(0.24)(+1.5)}{(590)(+0.0004)} = 1.53$$

* $H = -\beta(R_n + B)/(1 + \beta)$ and $E = -(R_n + B)/(1 + \beta)$.

† The derivation of Eq. (8-5) runs approximately as follows: Beginning with equations for logarithmic profiles of momentum and specific humidity—$((\partial U/\partial z) = u/kz$ and $(\partial Q/\partial z) = q/kz$—integrate these between heights z_1 and z_2. In the air layers far enough from the surface for turbulence to be fully developed (isotropic), the eddy velocities are equal in all directions; u may be substituted for w. With this substitution, combining the results of the two integrations with the flux equation $\xi = \rho \overline{qw}$ yields the desired result.

and

$$\xi = \frac{\rho k^2 (U_1 - U_2)(Q_1 - Q_2)}{[\ln(z_1/z_2)]^2}$$

$$= \frac{(10^{-3})(0.4)^2(-300 \times 60)(0.4)(10^{-3})}{(-2.0174)^2}$$

$$= -0.284 \times 10^{-3} \text{ gm cm}^{-2} \text{ min}^{-1}.$$

The evaporation rate calculated represents an energy transfer rate $E = L_v \xi$, of 0.167 cal cm^{-2} min^{-1}. According to the value of the Bowen ratio, then, the value of the sensible heat transfer rate, H, must be $H = \beta E = (1.53)(0.167) = 0.258$ cal cm^{-2} min^{-1}.

Table 8-3
SAMPLE OBSERVATIONS FOR CALCULATIONS OF H AND E

Level	Height (meters)	Temperature (°C)	Spec. Hum. (gm/gm)	Wind speed (cm/sec)
2	15	27.0	0.0123	1200
1	2	28.5	0.0127	900
Ratio	0.133	—	—	—
Difference	—	+1.5	+0.0004	−300

Q(gm/gm) = (6.5×10^{-4}) (vapor pressure in mb).
$c_p = 0.24$ cal gm^{-1} deg^{-1}.
$L_v = 590$ cal gm^{-1}.
$\rho = 0.001$ gm cm^{-3} is typical for air.

The literature on estimation of evaporation and on energy budgets of Type 1 systems is filled with many other methods, most of them elaborations of the ideas presented here. Why the overwhelming interest in these systems? The first impulses to evaluate energy budgets probably arose from a desire to find adequate explanations of the variety of fascinating observations in newly emerged microclimatology and from the knowledge that most of the energy driving macroscale atmospheric processes reaches the atmosphere through the variety of interfaces represented by Type 1 systems. Knowledge of the large scale behavior of the atmosphere, then, would seem to require knowledge of the behavior of the boundary conditions represented by its primary heat exchanger: the earth's surface.

Energy-Budget Type 1: Variability of Components

The variety of behavior found in Type 1 systems may be illustrated in several ways. Table 8-4 includes a variety of general comments about the differences found in different kinds of sites.

From the remarks in Table 8-4, one sees there is a sort of hierarchy of dissipation modes: *E* does most of the job so long as water is freely available at the surface, with *H* "taking up the slack" whenever water becomes less available. Except in moving water and granite-like materials, *B* is usually comparatively small.

Diurnal changes in the relative magnitudes of the terms in Eq. (8-1) are shown in Fig. 8-1 for a typical sunny day in mid-latitudes. The R_n term reflects the solar geometry by day and reaches a late-night negative equilibrium following a larger

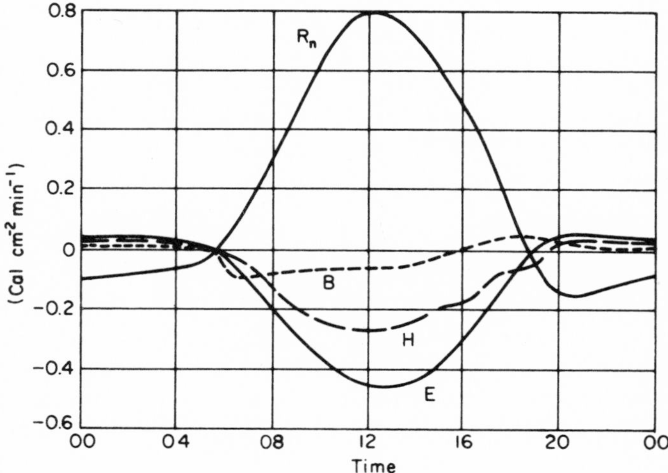

Fig. 8-1. Typical diurnal variations of the components of a Type 1 energy-budget system.

Table 8-4
SOME QUALITATIVE GENERALIZATIONS ABOUT THE ENERGY BUDGETS OF A VARIETY OF TYPE 1 SYSTEMS

Kind of site	Remarks on the energy budget of the site
Midlatitude vegetated areas generally	Dissipation modes in order of decreasing magnitude are *E*, *H*, and *B*.
Corn field or other lush growing crop	*E* as much as 80–85% of R_n.
Meadow	*E* largest dissipation mode during growth. *H* dominates when the grass is cut, and after it cures and mats to form an insulator, and a vapor barrier.
Sidewalk or stonefield	*E* is almost nil; *H* and *B* are about equal.
Desert	Similar to a stonefield, except *H* larger than *B*.
Ice cap (high latitude)	*E* and *B* almost nil; *H* and R_n are about equal and opposite in sign—*H* positive and R_n negative.
Open water	*E* and *B* completely dominate *H*; *E* is at the maximum possible given water and air temperatures; *B* large because of water's transparency and convective mixing of heat downward below the surface.
Still water	*B* is reduced as compared with open water, since convective mixing is reduced. As a result, the water surface becomes warmer, increasing both *E* and *H*.
Snow (mid-latitude)	Although snow is transparent to the incident sunlight, its large albedo keeps R_n comparatively small; as long as snow and air are below freezing, the excellent insulation of snow keeps *B* small, low vapor pressures keep *E* small, and *H* becomes a large fraction of R_n. Snow at the freezing temperature results in melting and mass transfer within the *B* term.
Leaf litter	The litter acts as a thermal insulator, keeping *B* very small; acts as a vapor barrier when dry, keeping *E* small; as a result *H* is much the largest mode of dissipation.

negative value just after sunset. This larger value is due to the larger upward flux of longwave radiation from the warmer surface, which cools through the night until the equilibrium is reached.

The E term is the largest dissipation term through the day, as noted in the comments on hierarchy above, and follows the R_n term closely, as is also true of the H term. Because of the abrupt changes in soil temperature profiles and heat flow at sunrise and sunset, as discussed in a previous chapter, the B term is asymmetric at these times. Unlike the E and H terms, the B term shows a daily net value of near zero: The heat stored in the soil by day is almost exactly the amount which flows from the soil at night.

In addition to the seasonal differences in the heat budget suggested for the meadow in Table 8-4, Fig. 8-2 shows month-to-month changes found in several of the world's climates. Here all terms are shown as positive in order to indicate more clearly the fractions of R_n represented by latent and sensible heat fluxes.[3] The soil heat term, B, does not appear in these monthly means because of the near-zero net values mentioned in connection with Fig. 8-1. In the four parts of Fig. 8-2, the change from a double to

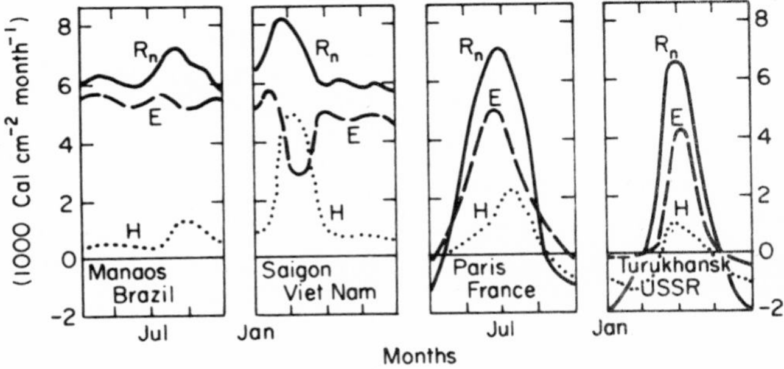

Fig. 8-2. Representative annual variations of the components of a Type 1 energy-budget system for four latitudinal and climatic combinations. (After Fig. 9 of Gates[3].)

a single annual cycle of sunshine in going from low to high latitudes shows clearly. The effect of cloudiness during the onshore monsoon of August–November in Saigon also shows clearly. At low-latitude stations, all terms have positive values all year, showing these areas to be the atmosphere's most dependable heat source. At higher latitudes both R_n and H become negative in winter, reflecting respectively the solar geometry of low sun angles and the poleward transfer of heat by the atmosphere's general circulation (refer to remarks under "Ice cap," Table 8-4). E is never negative on a net basis, regardless of latitude, at these stations.

Latitudinal effects on yearly net values show R_n decreasing poleward, becoming negative beyond the polar circles. E also decreases poleward, showing variations at any given latitude due to wet and dry climates. Except in hot deserts, E shows the net evaporative loss mentioned just above. Any change of latitude within desert climates produces changes in H nearly equal to the changes in R_n; that is, E changes scarcely at all.

Energy-Budget Type 1: Nature and Effects of the Soil Heat Component

Although net storage of heat in the soil may be quite small, under any circumstance the surface temperature is of utmost importance in heat dissipation. As a result, the effects of the nature of surface materials on the energy budget of a Type 1 system are very great indeed. Perhaps the most important characteristics of surface materials in determining the energy budget are the shortwave and longwave reflectivities. These determine the net heat load which must be dissipated, and the amount of radiative dissipation possible. Table 8-5 gives a brief list of natural surfaces and their typical reflectivities to demonstrate the range encountered in nature.

While the great variety of albedos is interesting and important, probably the most interesting information in the table is the fact that fresh snow, with the highest albedo, behaves almost as a black body with respect to longwave radiation. Water's transparency and a forest's internal reflections produce very dark appearances indeed, as any air traveler can testify.

In addition to differences in albedo for different materials, the albedo of the same material usually changes with the wetness of the surface. Because light energy is trapped by internal reflection

Table 8-5
TYPICAL VALUES OF LONGWAVE AND SHORTWAVE REFLECTIVITY FOR REPRESENTATIVE SURFACES

Surface	Albedo (shortwave)	Reflectivity (longwave)
Fresh snow	0.80	0.005
Dry sand (low sun)	0.60	0.05
(high sun)	0.35	0.05
Open water (low sun)	0.25	0.02
(high sun)	0.05	0.02
Typical fields	0.20	0.05
Coniferous forest	0.10	0.02

within the thin water film of a wetted surface, albedos are generally lower for wet surfaces, as shown in Table 8-6. An interesting question now arises. If a wet soil surface decreases the albedo, and

Table 8-6
THE EFFECT OF WETTING ON ALBEDO FOR VARIOUS MATERIALS

Surface	Albedo, dry	Albedo, wet
Gray sand	0.18	0.09
Matted grass	0.32	0.20
White sand	0.37	0.24

thus increases the absorptivity and the heat load, why are wetter soils usually cooler? In terms of Table 8-6, and the soil heat flow cases of Chapter 4, the paradoxical question is why an increase in albedo, such as from gray to white sand, will produce the same changes in soil heat storage and soil temperature profile as a decrease in albedo, such as from dry to wet sand. The answer is that such a large fraction of the heat load is dissipated through the E term in wet systems, the fraction going to the B term is on balance less, despite a slightly larger R_n.

As clearly indicated just above and in previous chapters, the nature of the subsurface materials strongly affects soil heat flow, which in turn affects surface temperatures and the entire energy budget for a Type 1 system. Examining the physical characteristics affecting the soil heat budgets of several materials will help to set this point straight. In Table 8-7 are given the thermal conductivities

Table 8-7
THERMAL PROPERTIES OF THE SOIL MATERIALS OF FIG. 8-3

Material	Thermal conductivity, k (Cal/deg/cm/sec)	Thermal capacity, ρc (Cal/deg/cm^3)
Granite	0.011	0.52
Sand heath	0.0008	0.35
Marshland	0.0015	0.70

and thermal capacities of the three types of subsurface materials depicted in Fig. 8-3. In the figure, the profiles are of daily maximum and minimum temperatures, not profiles representing any particular time of day. Several observations concerning the profiles are readily explained in terms of the thermal characteristics of the materials.

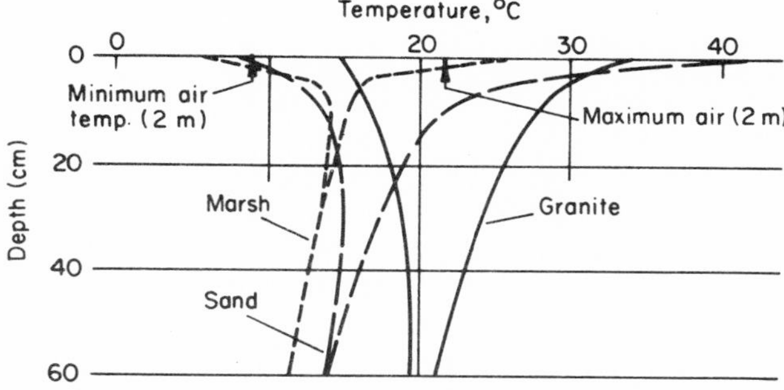

Fig. 8-3. Envelopes of temperature maxima and minima at depths in three types of soil material, together with air temperature maximum and minimum for the same day as the observations of soil temperatures. (After Fig. 74 of Geiger,[7] from data by Homén.)

First of all, the original observations from which these profiles come were all made at sites near one another in Finland, so we can assume the heat load on the system was essentially the same for each soil type. The soil heat-budget equation at midday (Eq. 4-6) shows sand, with conductivity and thermal capacity smaller than granite, to have lower temperatures at depth, higher surface temperatures, and high temperature gradients near the surface as compared with granite. These results become clear upon recalling the interpretation of the thermal characteristics: In granite the heat can flow downward from the surface more rapidly (higher k) and will raise a given volume of material to a lower temperature (higher ρc) than in sand.

A similar comparison of thermal characteristics between granite and the marshland would probably produce the conclusions that the marsh had higher near-surface temperature gradients (lower k), lower temperatures at depth (higher ρc and less outflow beneath the surface layers), and similar surface temperatures (a balance between a higher ρc and a lower k). Except for lower surface temperatures in the marsh, these conclusions are correct. Upon comparison of the marshland with the sand, we see again that conclusions based solely on conductivity and capacity can lead to trouble. Marsh has higher values of both characteristics than sand, and so should be expected to bear the same temperature relationships as granite does to sand: lower surface temperatures, lower near-surface temperature gradients, and higher temperatures at depth. Though surface conditions are as would be predicted, the temperatures at depth in marshland are the lowest of the three. We have come upon the effects of soil wetness mentioned above—the diversion of energy from the B term to the E term by evaporation from a wet surface.

The relative magnitudes of the B terms for the three soil systems may be ranked according to the diurnal temperature range at one of the lower depths, or more accurately according to the total area enclosed by the pairs of temperature profiles. By either ranking, the B terms in decreasing order are granite, sand, and marshland. Again, the marshland does not realize the large potential for heat storage to be expected from such a large thermal capacity because of the diversion of energy into the E term. This same diversion upsets any simple correlation between soil temperatures and the magnitude of the B term.

It has been implicit in these discussions that granite plays a very special kind of role in the energy budget of a Type 1 system. Because it can so quickly carry heat energy to great depths and can store it there with relatively small temperature increases, granite and granite-like materials produce the unusually large B terms noted in connection with the hierarchy of components mentioned under Table 8-4. The importance of this large storage capacity of granite for nocturnal conditions may be seen in Fig. 8-3 where the minimum temperature of the granite surface is higher than that of the air at 2 meters above the surface! The large flow of stored heat out of the granite at night simply prevented the surface from becoming the coolest stratum in the soil-air profile.

In summary of this section, it should by now be clear that the magnitude of the soil heat term, B, in the energy budget of a Type 1 system bears no direct and simple relationship to the soil surface temperatures in the system, which in turn have such a great influence on the other dissipation components of the system. The soil surface and subsurface materials play a major role in the determination of the energy budget, even though the soil heat flow may be only a small fraction of the total energy budget.

Energy-Budget Type 2: Estimation of Components

The working equation for this type of system typified by a single horizontal leaf, is

$$a_s(S_{di} + S_{ui}) + a_L(L_{di} + L_{ui}) + 2\varepsilon\sigma T_1{}^4 + (H_u + H_d) + (E_u + E_d) = 0. \qquad \textbf{(8-6)}$$

Although, as we have seen in Chapter 3, absorptivities for leaves vary according to wavelength, usable values of a_s for the shortwave spectrum as a whole have been obtained by careful measurements.[(4)] They lie in the neighborhood of 0.50 for most broadleaves.

Also by careful measurement, values of a_L for the longwave spectrum have been found to be very nearly 0.97 regardless of species.[5] Together with values for S and L obtained by appropriate radiometers,[6] these values permit calculation of the heat load on a given Type 2 system. Before leaving the radiant heat load on this system, the reader should note that while the shortwave absorptivity changes from 0.50 to 0.85 or so in going from an individual broad leaf to a forest, the longwave absorptivity changes only from about 0.97 to 0.98.

With the heat load components in the first line of Eq. (8-6) estimated, it remains to deal with the dissipation components in the second line. With thermocouple or appropriate longwave radiometer,[6] the leaf temperature may be measured and the back radiation estimated, noting $\varepsilon = a_L = 0.97$. As with the Type 1 system, it remains to partition the remainder of the energy budget between the E and H components. One may estimate the transpiration rate, ξ, with a form of potometer,[6] calculate the gross latent heat transfer as $(E_u + E_d) = E = L_v\xi$, and then consider the residual to be $H = (H_u + H_d)$. This is roughly equivalent to using a lysimeter for a Type 1 system.

Another method for partitioning energy between E and H is to make use of relationships developed in basic heat-transfer engineering, which enable estimation of H from wind speed and temperature data. In calm air, the situation known as *free convection* will produce currents of air moving vertically past the leaf because of density-buoyancy effects when the air and leaf are at different temperatures. With general wind movement, *forced convection*, the H term may likewise be estimated, and the E term considered a residual. The relationships for estimation of H in the cases of free and forced convection are

$$H_{\text{free}} = K_1(\Delta T/D)^{1/4}(\Delta T) = K_1(D)^{-1/4}(\Delta T)^{5/4} \qquad \textbf{(8-7)}$$

and

$$H_{\text{forced}} = K_2(V/D)^{1/2}(\Delta T) \qquad \textbf{(8-8)}$$

where K_1 and K_2 are constants which depend upon the physical units employed and the orientation of the leaf, ΔT is the temperature difference between leaf and air, V is the speed of the wind past the leaf, and D is a characteristic dimension of the leaf such as the width or the mean of the width and the midrib length.

As a sample calculation using Eq. (8-8), consider a leaf with dimension 10 cm in a wind of 220 cm/sec (about 5 mph). If the leaf is 5°C warmer (or cooler) than the air, and the value of the constant for these units and a horizontal orientation is 5.7×10^{-3},[6] then

$$H_{\text{forced}} = (5.7)(10^{-3})(220/10)^{1/2}(5) = 0.134 \text{ cal cm}^{-2} \text{ min}^{-1}.$$

If the leaf is warmer, of course, the heat is lost to the air. For all practical purposes, the same value of K as used here may be used in calculations of free convective transfer in calm air with Eq. (8-7). This method of estimating H and considering E as a residual has no counterpart among the methods mentioned for Type 1 systems.

Energy-Budget Type 2: Variability of Components

Knoerr and Gay[6] have presented leaf energy budgets for several situations and species, as summarized in Table 8-6. Although units for the environmental energy variables are expressed for a single cm^2 of leaf area, those of the energy budget are expressed for 2 cm^2 since all components have upward and downward fluxes combined. Several points are at once apparent from these energy-budget figures. First, leaf temperatures in sunlight typically rise 5 to 10°C above air temperature. Second, reradiation is by far the dominant mode of heat dissipation, as one would expect from its fourth-power relationship to leaf temperature. Third, considerable variation in E may be found among species. Fourth, E is not sensitive to wind speed beyond some very low speed. Fifth, despite no greater transpirational cooling with increasing wind speed, the leaf is nevertheless cooled by increased convection losses which rise in proportion to the square root of the speed. These increased losses by convection are simply transferred from the reradiation losses of cooler leaves. Finally, H is larger for smaller leaves, probably because their perimeter/area ratio is larger.

Table 8-8

COMPONENTS OF THE ENERGY BUDGETS OF THREE KINDS OF LEAVES[a]

Environmental conditions[b]:	S_{di}	S_{ui}	L_{di}	L_{ui}	Heat load	V
Free convection	1.07	0.24	0.89	0.90	2.38	0
Forced convection	1.06	0.10	0.63	1.04	2.20	50
Forced convection	1.06	0.10	0.63	1.04	2.20	425

Species	D	V	T_1	ΔT	Heat load	$2\varepsilon\sigma T_1{}^4$	H	E
Oak	9.8	0	50.5	9.2	+2.38	−1.72	−0.15	−0.51
	10.0	50	40.8	8.6	+2.20	−1.53	−0.37	−0.30
	10.0	425	37.6	4.1	+2.20	−1.47	−0.43	−0.30
Magnolia	10.0	0	60.5	19.2	+2.38	−1.96	−0.41	−0.01
	6.4	50	41.8	9.6	+2.20	−1.54	−0.45	−0.21
	6.4	425	39.1	5.6	+2.20	−1.50	−0.49	−0.21
Cherry	2.8	50	42.0	9.8	+2.20	−1.53	−0.57	−0.10
	2.8	425	38.4	4.9	+2.20	−1.48	−0.62	−0.10

[a] From Knoerr and Gay.[(6)]

[b] Units for variables: S and L = (cal/cm^2 min); V = (cm/sec); D = (cm); T_1 and ΔT = (°C); energy-budget components are in (cal/2 cm^2 min).

The heat load under forced convection, and the radiant dissipation from the second oak leaf were obtained as follows:

$$\begin{aligned}\text{Heat load} &= a_s(S_{di} + S_{ui}) + a_L(L_{di} + L_{ui})\\ &= (0.5)(1.06 + 0.10) + (0.97)(0.63 + 1.04)\\ &= 2.20 \text{ cal cm}^{-2} \text{ min}^{-1};\end{aligned}$$

and

$$\begin{aligned}\text{Radiant dissipation} &= 2\varepsilon\sigma T^4\\ &= (2)(0.97)(0.817 \times 10^{-10})(273 + 40.8)^4\\ &= -1.53 \text{ cal/(2 cm}^2\text{/min)}.\end{aligned}$$

Leaves in shade, as in the inner portions of a canopy, have an energy budget quite different from those in sun. Because of the greatly reduced heat load, dissipation can be accomplished at an equilibrium leaf temperature below air temperature. Under such circumstances, of course, H is positive as the leaf receives energy from the warmer air. The energy budget for the entire canopy, therefore, must be a weighted mean of the components of the shaded inner and sunlit outer leaves and branches of the vegetation. Also, account must be taken somehow of the various orientations of the leaves of the canopy. The results of Table 8-8 are for the somewhat artificial situation of a horizontal leaf.

Energy-Budget Type 3: Estimation of Components

As may be recalled from the laster chapter, if one neglects advective components, the principal adjustment in going from a Type 1 to a Type 3 system is the accounting necessary for the energy within the three-dimensional system. The working equation (Eq. (7-20) without the advective components) is

$$a_s S_{di} + a_L L_{di} + \varepsilon\sigma T_c^{\,4} + H + E + B + M_b + M_{Hp} + (M_H + M_E)_a = 0. \tag{8-9}$$

It is possible to estimate S, L, and the absorptivities by appropriate measurements with radiometers. While as noted above, the temperature of a leaf, T_1, may be estimated by means of either electrothermometry or radiation thermometry, obtaining an effective temperature for a three-dimensional canopy, T_c, would best be done by means of the longwave radiometer. Recalling that the streams of H and E in the lower (soil) hemisphere of the Type 3 system are by definition combined to form the B term, these three dissipative components may be estimated as in the case of a Type 1 system. Lysimetry would be pretty much restricted to canopies of relatively small plants, usually annuals. In using the schemes for partitioning energy between H and E by means of measurements of atmospheric parameters made at one or more levels above the system, care should be taken to position the instruments in the same relationship to the uppermost stratum of the canopy as they were to the soil or water surface in the Type 1 system.

Having estimated all the components of the vertical fluxes, then, it remains to estimate the several components of the energy stored and utilized within the canopy system. Suomi and Tanner[11] suggest that very carefully obtained measurements of temperature and vapor pressure within the canopy may be used to estimate three intracanopy terms by the following formulas:

$$M_{H\mathrm{p}} = \int_0^z \rho_\mathrm{c} c_\mathrm{c}(\partial T/\partial t)\, dz = \rho_\mathrm{c} c_\mathrm{c}(\Delta \bar{T})z, \tag{8-10}$$

$$M_{H\mathrm{a}} = \int_0^z \rho c_\mathrm{p}(\partial T/\partial t)\, dz = \rho c_\mathrm{p}(\Delta \bar{T})z, \tag{8-11}$$

$$M_{E\mathrm{a}} = \int_0^z (L_\mathrm{v} r/RT)(\partial e/\partial t)\, dz = (L_\mathrm{v} r/RT)(\Delta \bar{e})z \tag{8-12}$$

where ρc_p and $\rho_\mathrm{c} c_\mathrm{c}$ are thermal capacities of air and vegetation respectively, T is the temperature, t is the time, e is the vapor pressure, z is the depth of the canopy, L_v is the latent heat of vaporization, R is the specific gas constant, and r is the ratio of mole weights of water and air. Evaluation of these three terms leaves M_b, the energy involved in biological activities, as a residual. Even this term has been estimated by investigators employing carefully obtained measurements of carbon dioxide exchange within the canopy.

Energy-Budget Type 3: Variability of Components

As a general rule, the aggregate of all the M terms in Eq. (8-9) is only a very small fraction (perhaps 4–5%) of the energy budget of a Type 3 system involving air within the canopy.[7] The result is that the energy budget of such a Type 3 system is contained almost completely within the vertical flux terms, and upon evaluation appears very much like the budget for a Type 1 system in Fig. 8-1.[8] Careful examination of the vertical fluxes, however, reveals asymmetry which suggests something of the dynamic relationships within the canopy. The asymmetry may be seen in values of the Bowen ratio calculated from data obtained in a German spruce forest by Baumgartner (who is quoted at length by both Geiger[7] and Knoerr,[8]), as in Table 8-9. The midday minimum and the morning and afternoon maxima have been confirmed in studies of other canopies and are attributed among other things, to the relative effects of the linear increase in H and exponential increase in E with temperature of the canopy and canopy air.

Table 8-9
BOWEN RATIO FOR A SPRUCE FOREST CANOPY AS A FUNCTION OF TIME OF DAY[a]

Time	0700	0900	11–1300	1500	1700
Bowen ratio $= H/E$	0.15	0.80	0.50	0.60	0.30

[a] From Knoerr.[8]

Something of the seasonal variation in the energy budget suggested for a meadow in Table 8-4 may be seen in the summary of midday data from a Russian deciduous forest made by Knoerr: Table 8-10. The latent heat component is quite dominant during

Table 8-10
SEASONAL VARIATIONS IN THE ENERGY BUDGET OF A DECIDUOUS FOREST[a]

	Spring	Summer	Autumn	Winter
Net radiation (radiation load minus back radiation)	0.90	0.90	0.65	0.35
Latent heat flux,[b] E	0.62	0.65	0.35	0.10
Sensible heat flux, H	0.25	0.23	0.29	0.25
Bowen ratio, H/E	0.40	0.35	0.83	2.50

[a] From Knoerr.[8]
[b] Flux units are cal/cm^2 min. Soil heat flux is the small residual.

growth and full leaf, but the sensible heat component—which interestingly enough is almost constant with season—becomes relatively largest as leaves die and fall.

With one exception, the Bowen ratios in Table 8-8 are well above 1.0. That is, for individual, horizontal, sunlit leaves H is distinctly larger than E. Tables 8-9 and 8-10, on the other hand, show only one Bowen ratio larger than 1.0, and that for a leafless canopy. Thus, for entire canopies, E is generally larger than H in view of the weighted averaging of shaded and sunlit leaves.

The energy-budget equation for a Type 3 system consisting of vegetation growing in a shallow water body requires few modifications of Eq. (8-9). The temperature for back radiation would be that of the water surface, T_w, rather than a mean temperature of the three-dimensional upper canopy. The system, or intra-canopy, energy may be rewritten

$$M = M_b + M_{Hp} + M_{Hw}. \qquad \textbf{(8-13)}$$

Relative magnitudes among the various components here will doubtless be quite different from those in an air-filled canopy system. For one, the albedo of water at low sun angles is very high, though at high sun angles it is quite like the 0.15 or so of most vegetation. With a free water surface as the upper boundary of this system, E will be as large as possible with given conditions of wind and atmospheric moisture above the water surface. Complex instrumentation may often be avoided over water by estimating E as

$$E = CL_v U(e_w - e_a) \qquad \textbf{(8-14)}$$

where U is the mean wind speed measured at some rather low height over or near the water body, e_w is the saturation vapor pressure at the surface temperature T_w, e_a is the ambient vapor pressure at the site of U, and C is an empirical constant, whose value depends upon the positioning of instruments in the air, the size and shape of the water surface, etc. Because T_w will most likely be lower than T_c for a nearby air-filled canopy of comparable depth, the radiative and sensible heat dissipation components will be smaller. Finally, because the thermal capacity of water is roughly 4000 times that of air, any significant rise in water temperature in the canopy will represent a much greater energy change than in air. Since the thermal capacity of water is 1.0, and since in Eq. (8-13) M_{Hw} is likely to overshadow the other two terms completely, M may be estimated by rather simple measurements of water temperature as $(\Delta \overline{T})z$, according to Eq. (8-11).

Equation (8-14) is usually employed when the evaporation rate, rather than the complete energy budget, is the object of study. For a complete energy budget, if one must rely on the simple instrumentation required for Eq. (8-14), B may either be estimated by measurements of temperature at the bottom of the pond or may be assumed zero. In either case, H will be the residual of the other terms.

As with the case of an air-filled canopy, the density of vegetation in a pond system will have an effect on the mobility of the canopy medium, and thus will affect most of the components of the energy budget for the system.

Energy-Budget Type 4: Estimation of Components

The equation for the energy budget of a Type 4 system—Eq. (7-21)—is deceptively simple. The A's, or areas of transfer surface appropriate to the various components, are often simple in concept and complex in fact. For example, the surface area involved in radiant transfer of shortwave energy, A_s, is different for the direct solar beam than it is for scattered sky radiation. This problem was of no concern in systems of Type 1, 2, and 3, since there only unit horizontal areas were considered.

We may borrow from the ideas of Gates[9] in defining the areas of transfer for a Type 4 system of complex shape. To begin, we may separate the direct-beam and extended-source areas of A_s into A_{sd} and A_{se}, respectively. Here A_{sd} is the area of the shadow cast by the system on a plane perpendicular to the solar beam and may be estimated by measuring a shadow area for different orientations. A_{se}, on the other hand, is the area "seen" by the entire spherical surroundings of the system. Since portions of the system's surface—whether a building, a tree, or an animal—overlap one another as seen from the surrounding sphere, A_{se} will generally be less than the total surface area of the system, A, and will even be a function of the posture of the system in the case of an animal. Gates defined a fourth area, A_c, as being that area of the system in contact with the solid substratum. For a man standing, it is the contact area of his feet; for a house without basement, the floor area; and so on. A_c is, like A_{se}, a function of posture.

For Type 4 systems which are geometrically comparatively simple, such as a building, the four areas may be calculated. For slightly more complex shapes, such as a man standing with arms at his side, rough estimates of the various areas may be made by considering the system to be a vertical cylinder. As pointed out by Lee,[10] for example, such a cylinder would have $A_{se} \doteq \pi Dh$ and $A_{sd} \doteq Dh \cos \theta$, where D is a representative "diameter" of the man, h his height, and θ the altitude angle of the sun.

An extension of this cylinder model for a quadruped, for instance, would lead to $A_{se} \doteq \pi Dl$ and $A_{sd} \doteq Dl \sin \theta$, where l is the length of the animal and the other symbols are as before. These estimates are only very approximate, of course, but have the merits of being simple and direct. Greater precision, one might argue, would be warranted only in very careful analyses where such precision is well assured in other variables of the problem.

Gates[9] described another approach in which precision is sought for systems of very complex shape which are also capable of being represented by exact scale models. He estimated the total areas of conifer branches and of a lizard, as examples of these very complex shapes, by an electrolytic technique in which a metal casting of the system is used as one electrode, and the current flowing to the other electrode is a function of the total area, A, for a given voltage. In the case of the lizard, Gates obtained the area in contact with the substratum, A_c, by covering that portion with an insulating material, then using the casting again as an electrode and noting the smaller current.

The area A_{se} may be estimated by suspending the casting in an evacuated chamber whose inner wall temperature is controlled by a surrounding flow of water. By electrically preheating the suspended casting, and then recording its temperature as it cools by radiative transfer to the chamber walls, the area of the casting "seen" by the chamber walls, A_e, is estimated by

$$A_e = [mc_c(dT_c/dt)]/[\varepsilon_c \sigma(T_c^4 - T_w^4)] \qquad \textbf{(8-15)}$$

where m is the mass of the casting; c_c, T_c, ε_c are the specific heat, temperature, and emissivity of the casting; and T_w is the wall temperature. This equation is valid because the numerator expresses the rate at which the heat content of the casting is changing while the denominator expresses the rate at which heat energy is being transferred from the casting to the walls by radiation. The vacuum in the chamber assures that only radiant transfer is involved in the cooling.

Having estimated A_e, Gates uses a value for A_{se} of

$$(A_e/A)(A - A_c),$$

since the area insulated electrically to obtain A_c will not be insulated thermally in obtaining A_e. For a lizard (A_c/A) was about 0.15. For a standing man, it would probably be less, and for a reclining man much larger. For buildings, it might range from something like 0.10 for skyscrapers to 0.30 for a ranch-style home. For a tree or a tree branch, this area of solid supporting contact, A_c, would probably be only a small fraction of A. Thus, the attention given to estimation of A_c and to allowing for it in estimating A_{se} depends to some extent on the shape and posture of the system under study. Relative areas which Gates found in working with conifers were $(A_{se}/A) = 0.94$ for fir and 0.88 for spruce, and $(A_{sd}/A) = 0.32$ for fir and 0.34 for spruce with the branch perpendicular to the solar beam. This result for the cylinder-like conifer branchlets, $A_{sd} = 0.33\ A$, closely resembles the result one might get from the cylinder model in which $A_{sd} = (1/\pi)A$ for a perpendicular solar beam.

Having now somehow estimated the areas A, A_{sd}, A_{se}, and

A_c, we may rewrite Eq. (7-21) as follows:

$$a_s S_d A_{sd} + a_s S_e A_{se} + a_L L_i A_{se} + \varepsilon\sigma A_{se} T_s^4$$
$$+ h_c(A - A_c)(\Delta T) + kA_c(dT/dz) + E + M = 0. \qquad \textbf{(8-16)}$$

Here S_d is the flux of direct solar radiation,* S_e is the flux of scattered and reflected shortwave energy from the surrounding sphere, L_i is the flux of longwave energy from the surrounding sphere, and T_s is the effective surface temperature of the system under study. The first line of the equation, then, is the net radiant flux, including the dissipative back radiation.

The H term is divided into convective and conductive components in the first two terms of the second line. In a manner similar to that suggested in Eqs. (8-7) and (8-8), the convective losses from the complex surface configuration of a Type 2 or a Type 4 system may be estimated by means of the product of a convective transfer coefficient, h_c, and the temperature difference between the system surface and the surrounding air, ΔT. For a cylinder-like system, the experimentally determined h_c bears a proportional relationship to the $\frac{1}{3}$ power of the wind speed, rather than the $\frac{1}{2}$ power as in the case of Eq. (8-8) for a flat-plate system. The conduction component in Eq. (8-16) involves the thermal conductivity, k, of the material upon which the system rests, and the temperature gradient within that material next to its area of contact with the system A_c. This conductive component may be likened to the B term in Type 1 and 3 systems.

The dissipation of energy by latent heat, E, and the stored and utilized energy, M, must be obtained for the system as a whole. E may be estimated by weight loss during a unit of time, which would be analogous to a lysimeter estimation, or by gas exchange methods comparing moisture content of air arriving within a test chamber to air leaving the chamber containing the entire system.

* S_d must be measured in the same plane as A_{sd} for the relationship given in the equation to hold. If both are measured in the horizontal plane, for example, rather than in the plane perpendicular to the solar beam, the product $S_d A_{sd}$ would give the same result.

Table 8-11
CALCULATION OF THE ENERGY BUDGET OF A SMALL ANIMAL AS A TYPE 4 SYSTEM

Basic Variables: The environment:	The animal:
$S_d = 1.60$ cal cm^{-2} min^{-1}	$a_s = 0.90 \quad a_L = \varepsilon = 0.97$
$S_e = 0.10$ cal cm^{-2} min^{-1}	$A = 75$ cm^2; $A_{se} = 60$ cm^2
$L_i = 0.25$ cal cm^{-2} min^{-1}	$A_{sd} = 18$ cm^2; $A_c = 10$ cm^2
$T_{air} = 27°C = 300°K$	$T_s = 31°C = 304°K$
$k = 0.005$ cal cm^{-1} deg^{-1} min^{-1}	$h_c = (5 \times 10^{-3})V^{1/3}$ cal cm^{-2} min^{-1} deg^{-1}
$dT/dz = 20°$ cm^{-1}	when V is in cm sec^{-1}
$V = 27$ cm sec^{-1}	$E = 0.50$ cal min^{-1}
$\sigma = (0.817 \times 10^{-10})$ cal cm^{-2} min^{-1} °K^{-4}	$M = 0.90$ cal min^{-1}

Calculations:	
Direct solar radiation load $= a_s S_d A_{sd} = (0.9)(1.6)(18)$	$= +25.9$ cal min^{-1}
Indirect shortwave radiation $= a_s S_e A_{se} = (0.9)(0.1)(60)$	$= +\ 5.4$ cal min^{-1}
Longwave radiation load $= a_L L_i A_{se} = (0.97)(0.25)(60)$	$= +14.6$ cal min^{-1}
Metabolic heat production $= M$	$= +\ 0.9$ cal min^{-1}
Longwave radiation from integument $= \varepsilon\sigma A_{se} T_{es} = (0.97)(0.817 \times 10^{-10})(60)(304)^4$	$= -40.6$ cal min^{-1}
Convective sensible heat loss $= h_c(A - A_c)(T_{air} - T_s) = (5 \times 10^{-3})(27)^{1/3}(65)(4)$	$= -\ 4.9$ cal min^{-1}
Conductive sensible loss to substratum $= kA_c(dT/dz) = (0.005)(10)(20)$	$= -\ 1.0$ cal min^{-1}
Latent heat loss from body wet tissue	$= -\ 0.5$ cal min^{-1}
	0.0 cal min^{-1}

M may also be estimated by gas exchange techniques involving for example carbon dioxide; or both may be considered simply the residual of the other terms of the energy budget.

Table 8-11 includes representative values and calculations for the energy budget of a small animal such as the lizard described by Gates. The budget obtained is, of course, the response of the system to a single set of environmental parameters. Responses of a man, representing a Type 4 system, to variable environmental conditions are described in a later chapter.

Concluding Remarks

As noted at the beginning of Chapter 7, detailed research on the energy budget of a particular system will not provide information on all important ecological interactions with the environment, but will shed considerable light on energy flows between system and environment. Such knowledge, combined with knowledge of, for example, thermoperiodic effects, may lead the way to effective artificial control of a system's environment to reach some stated objective of system management. Likewise, such a combination of knowledge may serve to identify and explain behavioral responses of organisms to the environment orboth. Finally, thorough understanding of and complete information on the energy-budget concept forms the basis for artificial controls on man's environment, from ordinary earth-bound architecture to design and operation of closed-system environments such as space capsules, fallout shelters, and even the planetary architecture of the future.

In the chapters to follow, then, we shall examine some of the other types of ecological interactions of organisms with their environments, and some of the means employed in artificial control of those environments. First, however, it will be well to examine the ground over which we have traveled up to this point. Some thoughts about instruments and data collection, topics only implicit before now, will serve as the framework for the résumé.

REFERENCES

1. See e.g., E. K. Webb, "Aerial Microclimate," *Meteor. Monograph* **28**, Amer. Meteorol. Soc., Boston, Massachusetts, 1965.
2. C. W. Thornthwaite and B. Holzman, *Monthly Weather Rev.* **67**, 4 (1939).
3. D. M. Gates, "Energy Exchange in the Biosphere," Harper, New York, 1962.
4. R. Birkebak and R. Birkebak, *Ecology* **45**, 646–649 (1964).
5. D. M. Gates, *Appl. Optics* **4**, 11–20 (1965).
6. K. Knoerr and L. W. Gay, *Ecology* **46**, 17–24 (1965).
7. For discussion of these intracanopy terms, see e.g., R. Geiger, "The Climate near the Ground," p. 310, (4th ed.), Harvard Univ. Press, Cambridge, Massachusetts, 1965.
8. See, e.g., K. R. Knoerr, "Partitioning of the Radiant Heat Load by Forest Stands," *Proc. Soc. Amer. Foresters*, pp. 105–109 (1964).
9. D. M. Gates, *Amer. J. Botany* **51**, 529–538 (1964) and *Ecology 48*, 315 (1967).
10. D. H. K. Lee, and J. A. Vaughan, *Intern. J. Biomet.* **8** (1), 61 (1964),
11. V. Suomi and C. B. Tanner, *Trans. Amer. Geophys. Union* **39**, 298–304 (1958).

PROBLEMS

8-1. Both the radiant and the sensible modes of heat dissipation by a surface are temperature sensitive. A certain dry surface (longwave emissivity = 0.95) dissipates by convection 0.10 cal cm^{-2} min^{-1} with a certain wind speed and a 5°C temperature difference between itself and the free air nearby. Beginning with a surface temperature of 20°C, and assuming a constant air temperature of 20°C, calculate and plot the following three variables as functions of increasing surface-air temperature difference:

(a) energy dissipated as L_{uoe} (cal cm^{-2} min^{-1}),
(b) energy dissipated as H_{uo}, and
(c) the ratio H_{uo}/L_{uoe}.

8-2. Describe an experimental setup of the radiation instruments in Table 8-2 which would provide the means for calculating the six radiation streams of a Type 2 system: S_{di}, S_{ui}, L_{di}, L_{ui}, L_{uoe}, and L_{doe}.

8-3. A solarimeter measures a downward shortwave flux of 1.40 cal cm^{-2} min^{-1}. Calculate the upward shortwave fluxes from surfaces of dry white sand and wet matted grass and from a coniferous forest stand beneath the solarimeter.

8-4. Estimate the fraction of potential soil heat flux at the marsh surface in Fig. 8-3 which was diverted to evaporative cooling during the day of observation. Assume R_n and H were the same for granite and marsh, and that there was no evaporation from the granite.

8-5. If a leaf of characteristic dimension 10 cm lost heat convectively at a rate of 0.20 cal cm^{-2} min^{-1} in a wind of 200 cm sec^{-1}, what would be its convective heat loss rate in calm air under the same temperature conditions?

8-6. The water in a swimming pool 10 meters square and 1 meter deep was heated 1°C by the sun during two midday hours. What was the average flux of solar radiation (cal cm^{-2} min^{-1}) during the two hours if half the energy was diverted to evaporative cooling? Neglect changes in water mass due to evaporation. How much evaporated water did you just neglect?

8-7. If the small animal in Table 8-11 had a surface which had a zero albedo, calculate:

(a) the resulting increase in its heat load, and

(b) the resulting increase in its surface temperature at equilibrium, all other factors remaining unchanged.

8-8. You are interested in obtaining the radiative surface area of a piece of granite by observing its cooling rate in an evacuated chamber as suggested by Gates. You know the stone has a mass of 5 kg and that for longwave radiation it is essentially a black body. The chamber's wall temperature is controlled at 20°C and also acts as a black body. If the cooling rate of the stone is observed to be 2°C/min when it has a temperature of 85°C, what is its radiative surface area?

Chapter 9

INSTRUMENTATION AND DATA COLLECTION

IN PREVIOUS DISCUSSION, emphasis has been on the physical environment, although consideration of organisms has been included here and there. This chapter is the watershed between emphasis on environment and emphasis on organism.

We have managed to consider rather completely the nature and behavior of the physical environment without ever directly confronting problems of instrumentation and data collection. This book is not primarily concerned with instrumentation, but there is time during this pause for some general thoughts about the rationale with which the biometeorologist approaches his task of data gathering.

Physical Environment: a Synopsis

Recall that the study of the physical environment has been keyed to a study of the primary heat transfer processes: radiation, conduction, convection-advection, and latent heat flow. We have come to know that one needs two kinds of information to evaluate heat transfer in the environment. First, one needs data which describe the tangible elements of the environment, such as information about absorptivity, thermal conductivity, mass, distance, and time.

Second, one needs data which describe the momentary energy status of the tangible elements, such as data about radiant flux, temperature, wind speed and direction, and moisture content. In describing an environment, a biometeorologist sometimes uses one of these measures directly, such as temperature. At other times, he uses synthetic measures, such as evaporation rate, which combine data of the first and second kind.

The Objective of It All

No matter how the investigator chooses to describe the environment, with direct or with synthetic measures, he will do well to keep in mind the objective of his work in biometeorology. As described in Chapter 2, the objective is to identify aspects of life which are responses to the atmospheric environment and to explain the responses identified. Thus, describing the environment is not the objective itself. Describing the environment is only a means to reaching the objective. Explaining things about organisms is the objective. In this book, the organismic responses for which explanations are offered fall under headings such as growth and development, reproduction, movement and other behavior, and death.

In Chapter 1, mention was made of two basically different approaches to experimental biometeorology. The first examines organismic reactions to a prescribed and controlled set of environmental variables. The second seeks the combination of environmental variables responsible for an observed reaction. In either case, an investigator requires a description of the environment sufficient for his research plan. It is the nature of this "sufficient" description which we propose to examine now in broad terms. In addition, we will have a look at the means for obtaining the sufficient description through the use of instruments.

Basic Requirements for Proper Data Collection

It has been proposed[(1)] that there are four basic requirements for data collection leading to a sufficient description of the physical environment in biometeorological research. They are (a) knowledge of the nature of the environment, (b) a scheme for characterizing the environment, (c) knowledge of instruments available and of their interactions with the environment, and (d) a plan for obtaining measurements compatible with the scheme for characterization. Examine the implications of each of these in turn.

Knowledge of the Nature of the Environment

Underlying either of the two basic approaches to biometeorological research is a hypothesis about a mechanism linking organism and environment. If for no other reason, an investigator needs to know the workings of the environment in order to make rational adjustments of environmental variables during the process of refining his hypothesis. The reader should, at this point, have a reasonably good grasp of environmental behavior from having read and understood the first half of this book.

A Scheme for Characterizing the Environment

In refining his working hypothesis, the biometeorologist tries and discards a series of combinations of environmental variables as a result of comparisons with data on the response of his organism. Each combination constitutes a scheme for characterizing the environment. Some schemes may be simple and may involve only a statement about the presence or absence of some element in the environment. Others may range upward in complexity through statements about the heat or moisture content of air, and about the rate of change or rate of transfer of some constituent, to more sophisticated schemes.

The basic scheme employed in this book is the energy budget. In later chapters the reader will find instances where simpler schemes are used in examples taken from the literature of biometeorology, but in large measure the energy budget underlies these schemes for characterizing the environment. The basis for selection of the energy budget and for the continual reference to it is the proposition put in Chapter 2 that environmental energy underlies life itself and that the various forms and actions of organisms constitute responses to variations in the processes of energy transfer.

There are doubtless as many ways to characterize the physical environment as there are organismic reactions to explain. How does an investigator know which to choose? How does he know when a description of the environment is sufficient for his purposes so that he may end the process of trying and discarding?

The most successful schemes have certain features in common.[(1)] The list may be used to check whatever is chosen by a biometeorologist. The features proposed are:

(a) an accurate reflection of the working hypothesis;
(b) quantification based on physical laws and principles, as opposed to a basis of arbitrary or subjective measure scales;
(c) placement of the environment described in the spectrum of possible environments;
(d) internal consistency, in the sense that, for example, events mutually exclusive in nature are kept so in the analysis;
(e) compatibility, as for example the expression of all variables involving energy in units of energy; and
(f) provision for handling all reasonably pertinent factors.

How does the energy-budget scheme compare with this list? It is an accurate reflection of the fundamental proposition about energy in the environment underlying life. It is certainly quantified physically, and the total amount of energy budgeted at any moment places the system in a spectrum. Internal consistency and compatibility are present, and the provision for metabolic, or biological energy extends the scheme to all variables reasonably pertinent to the problems confronted by the biometeorologist.

Knowledge of Instruments and Their Interactions

With consideration of this requirement for proper data collection, we get to the heart of this chapter. There is a point of view which says "theories may come and go, but a good measurement will live forever." The statement has merit if one first agrees that to be good a measurement must truly represent what it purports to represent. Said another way, a measurement is good if the user knows what it represents and if it fits his scheme. This knowledge comes with knowledge of available instruments, their advantages, and their disadvantages.

As with the tools of any craft, those of the biometeorological instruments specialist are extremely varied: some flexible and multipurpose and others very specialized.* And as with the craftsman, the biometeorologist who uses the correct tools in the proper manner is the one who in general does the best work. The right instrument correctly used avoids errors which are avoidable and takes account of those which cannot be avoided.

These statements just made tend to be platitudes. What are some substantive examples of incorrect use to help make the points under discussion?

In the advertisements for and the technical literature accompanying the Eppley pyrheliometer, it states quite plainly that this instrument is for the measurement of solar radiation.(2) The fact that solar radiation and sky radiation (S_d and S_e) are predominantly short wavelengths permits the use of a glass canopy on this instrument to provide a weather shield and to reduce convective cooling of the sensitive elements. Despite these explicit statements by the manufacturer, one hears and reads of biometeorologists using the Eppley to "measure" the all-wave radiant energy flux in a "biotron" where artificial light is used to simulate the natural incoming radiation out-of-doors.

An example of interaction between instrument and environment is the often discussed disruption of the normal wind flow by a bulky precipitation gauge. Another is the obvious interference with normal heat and moisture flow by the introduction of a lysimeter in a study site.

The point is that an investigator who uses instruments should be aware of the errors which go with some instruments and which may be avoided with others. He can then make an informed

* Several major references on available instrumentation appear in the Supplementary Reading List at the end of this book.

choice and allow for the errors of his chosen instrumentation system in other ways.

Even though the biometeorologist may be well aware of the theory of his instruments and the types of problem for which they are appropriate, he will still have to contend often with errors which are unavoidable in any instrument system he may choose. Table 9-1 summarizes a list of types of inherent error in terms of the functioning of an ordinary mercury-in-glass thermometer.

To appreciate the contents of Table 9-1, the reader should recall the discussion in Chapter 6 in which the variable and the quasiconstant components of an environmental variable are considered. For temperature, the two components would be combined in an instantaneous value T' as

$$T' = T + t,$$

where T represents the slowly changing (low frequency) components and t the rapidly fluctating (high frequency) components. This convenient simplification, of course, masks the fact that the two classes of variability represented by T and t are in fact each made up of a spectrum of frequencies, such as a band of frequencies in radiation. In terms of the thermometer of Table 9-1, then, one is likely to look at the registration and read the value of T, the slowly changing component, since the nature of the instrument suppresses the high frequency contribution of t. If the observer is interested in T rather than in T', all is well. If he is, however, interested in T', the true temperature at the moment of looking, he should be aware of the resolution error of the thermometer.

Plan for Obtaining Compatible Measurements

Once instrumentation and a plan for use of the raw data have been settled on, it is necessary to have a plan for obtaining data of a proper nature for calculating the necessary factors or statistics. Obvious as this may seem, the point is often overlooked in the planning stage. For example, sample sizes must be such as to

Table 9-1

TYPES OF INHERENT INSTRUMENT ERROR IN TERMS OF THE MERCURY-IN-GLASS THERMOMETER

Type	General description	In terms of the thermometer
Resolution	Reproduction of one frequency band out of true proportion to another frequency band.	Reproduces diurnal variations but not those of a few seconds' duration, however large.
Phase distortion	Reproduction of one frequency band out of true sequential relationship to another.	None apparent.
Linearity	Lack of simple proportionality between signal and output.	"Necked down" portion of bore with a linear etched scale.
Sensitivity	Insufficient or variable proportion of change in output relative to the amount of change in signal.	Etched marks too wide for the distance between them, or bore diameter too large for the capacity of the mercury reservoir.
Dead zone	Static friction too great between moving parts.	Surface attraction between mercury and glass.
Noise	Output due to extraneous environmental variations combined with those sought.	Wind on the bulb causing pressure oscillations inside.
Calibration	Using incorrect statement of relationship between signal and output.	Etched marks and numbers "slipped" to one side during manufacture.
Exposure	Improper or inappropriate installation.	Improper ventillation or shading.
Sampling	Location or observation times not representative.	Use of one or two readings per day to estimate diurnal march of temperature.

exercise proper control over environmental variability; instruments must be exposed at several levels if gradients figure in subsequent calculations; and observation schedules must be realistic in terms of actual data requirements on the one hand and logistics on the other.

Concluding Remarks

The biometeorologist approaching the problem of data collection must have clearly in mind the objective of his effort. With this objective as his guide, the quality of his efforts should be in direct proportion to the amount of thoughtful planning which goes into design of an instrument system, a plan for data acquisition, and a plan for analysis of the data acquired through the instruments.

The aim, on balance, is to obtain the right numbers of observations of the environment, of the right precision, and of the right kind to explain the organism-environment interactions specified in the research objective. The reader is invited to use the remarks of this chapter as a standard against which to measure the methods and conclusions discussed in the chapters to follow. It is the intention in what follows to present some of the results of biometeorological research which, to repeat, are in the form of descriptions and explanations of various organismic responses to the physical (especially atmospheric) environment.

REFERENCES

1. W. P. Lowry, Biometeorological data collection, *Phytopathology* **53**, 1200–1202 (1963).
2. An example of advertising for the Eppley may be seen in *Bull. Amer. Meteorol. Soc.* **47**, 64 (1966). A description of its construction and of the theory of its operation is in W. D. Sellers, "Physical Climatology," p. 73, Univ. of Chicago Press, Chicago, Illinois, 1965.

PROBLEMS

9-1. You are to use the Type 3 model to estimate the energy budget of a corn field. Specify (a) the kinds of measurements to be made, (b) where they are to be made, and (c) when they are to be made.

Then specify (d) the way in which each component of the budget is to be obtained from the raw data. Have you specified observations not used in obtaining the complete budget? Have you omitted specifications for data required?

9-2. If you had to have each term of your budget in Problem 9-1 within at most $\pm 20\%$ of true value, estimate the percentage error you could allow in each of the measurements you have prescribed.

9-3. You intend to obtain a space-averaged value of the B term in an energy budget by using temperature profile data from a number of locations simultaneously. You estimate the area to be sampled is 70% in full sun and 30% in full shade at any given moment. Also, you estimate the values of B at sunny locations will vary $\pm 10\%$ about their mean and at shaded locations, $\pm 5\%$. If you need to be sure of your space-averaged B term within $\pm 2\%$, how many sampling points (soil temperature profiles) will you need?

SECTION III

The Biological Environment

Chapter 10

PLANTS AND THE ATMOSPHERE: PHYSIOLOGY

Introduction

WHILE IN EARLIER CHAPTERS we have considered plants as radiating materials and as components of energy-budget systems, in this chapter the emphasis will be on some of the details of atmospheric effects on the physiology of plants. In particular, we will be concerned with the processes of photosynthesis (PSN), respiration (RESP), and transpiration (TRP). While photoperiodism is certainly environment-sensitive, it is governed by more regular and predictable variations in the environment than those of primary concern in this book. Thermoperiodism, however, we shall consider in the next chapter.

The core of plant interactions with the atmospheric environment lies in the gross equations for photosynthesis:

$$CO_2 + 2H_2O^* \xrightarrow[\text{green plant}]{\text{light}} (CH_2O) + O_2^* + H_2O, \quad \textbf{(10-1)}$$

and respiration:

$$(CHO) + O_2 \xrightarrow[\text{organism}]{\text{living}} CO_2 \times H_2O. \quad \textbf{(10-2)}$$

In PSN, the elements of two atmospheric gases are combined with light energy captured by the chloroplasts of green plants to form plant materials and a third atmospheric gas. The asterisks in Eq. (10-1) signify that the oxygen released in molecular form came originally from water vapor rather than from carbon dioxide. In organismic respiration, plant or animal materials are burned to form the two gases from which plant materials were originally formed. For simplification, the product of PSN is written as a sugar, and fuel materials for RESP are written as a carbohydrate.

In Fig. 10-1 an elementary model of plant leaf anatomy shows

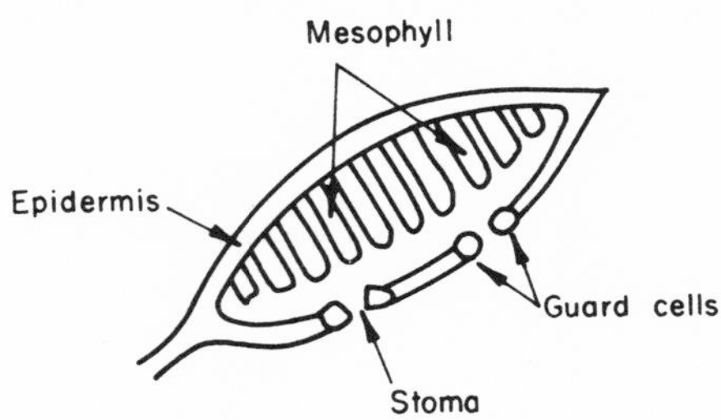

Fig. 10-1. Schematic of the essential plant parts involved in the discussions of this chapter.

the essential leaf parts involved in PSN, RESP, and TRP. The gaseous constituents of Eqs. (10-1) and (10-2) pass back and forth through the stomata, each according to the gradient of its own partial pressure. Except for the ports provided by the stomata, whose cross-sectional areas are each controlled by a pair of guard cells, the epidermis is essentially impervious to atmospheric gases. Within the leaf, the constantly moist surfaces of the mesophyll, which by their structure provide an astoundingly large area for gas exchange within an individual leaf, provide the loci for the reactions of PSN. As suggested in the figure, stomata are found mostly on the undersides of leaves in most plants.

PSN, of course, proceeds only in light, while RESP goes on day and night. RESP consumes some of the products of PSN, so that for a plant to grow, there must be an excess of the products of photosynthesis above the amount of plant materials consumed in respiration. The rate of accumulation of plant materials in a growing plant may be related to net, or apparent, photosynthesis:

$$\text{Total PSN} - \text{RESP} = \text{Net PSN}. \qquad \textbf{(10-3)}$$

While it is difficult to measure Total PSN, several methods are available for estimating Net PSN.

A qualitative technique known as a "starch test" may be used to demonstrate that plant materials of some magnitude have been deposited at a certain site during the course of the test. Increases in dry weight of plant materials during a period of time provide a rudimentary quantitative estimate of the rate of Net PSN, but methods for precise monitoring of gas concentrations may be used for nearly instantaneous estimates of the rate of Net PSN. For example, a knowledge of the concentrations of CO_2 in the air streams entering and leaving an enclosed chamber in which a living plant is engaged in PSN and RESP permits calculations of Net PSN moment by moment. Carbon dioxide is preferred as the gas to be analyzed because its concentrations are lower than those of oxygen; thus CO_2 measurements provide greater percentage variations and more accurate results.

The amount of PSN which takes place on earth may be suggested by estimates of the number of tons of carbon fixed each year in the process: 19×10^{10} tons C/year. Of this amount about 90% is fixed by aquatics and the remainder by land plants. Forests alone fix about 7%, leaving only about 3% for all the managed and unmanaged fields on earth.

Photosynthesis

While Eq. (10-1) gives the overall picture of the nature of PSN, the impact of the atmospheric environment may be better appreciated by separation into two equations:

$$H_2O + \text{Carrier} \xrightarrow{\text{light}} O_2 + \text{Carrier} \cdot H^+, \qquad \textbf{(10-4)}$$

$$\text{Carrier} \cdot H^+ + CO_2 \longrightarrow (CH_2O) + \text{Carrier}. \qquad \textbf{(10-5)}$$

Equation (10-4) is known as the "light reaction," in which light energy splits water molecules and attaches hydrogen ions to so-called carrier molecules. These reduced carriers enter into a "dark reaction" with carbon dioxide—Eq. (10-5)—to form the building blocks of plant materials. Carrier molecules shuttle back and forth between the two processes in such a way that the processes proceed in concert to form plant materials.

The light reaction is photochemical, and as such its rate is sensitive to the rate at which light energy arrives and is made available. Its rate is insensitive, however, to temperature. The dark reaction, on the other hand, is not photochemical, is insensitive to light, and is sensitive to temperature. Given a sufficient pool of carrier molecules (water for PSN is essentially limitless in a healthy leaf), the light reaction proceeds at a rate proportional to the arrival rate of light energy, producing molecular oxygen and reduced carriers. If for some reason the light reaction proceeds more rapidly than the dark, the pool of oxidized carriers becomes depleted and insufficient for further increase in the rate of the light reaction.

Either low temperatures or a low concentration of CO_2 might cause the reduced rate of the dark reaction, and thus of PSN, even in the presence of sufficient light. PSN as a whole, then, may be heat-limited or CO_2-limited. Clearly, even in warm temperatures and a CO_2-rich atmosphere, PSN may proceed slowly under low light intensities. In this light-limited condition, PSN is slowed by a drop in the availability of reduced carriers for the dark reaction. Thus, Total PSN in a leaf proceeds overall at a rate limited by one of three factors in the atmospheric environment: heat, light, or carbon dioxide. As will be seen presently, factors of the soil environment also affect the rate of total PSN.

Low temperature, as just noted, may limit the rate of total PSN by limiting the rate of the dark reaction. While theoretically the rate of Total PSN may not be limited by high temperature, in fact, excessive heat may cause damage to plant tissues and a resulting reduction in the supply of water for the light reaction. More commonly, effects of high temperature are referred to the rate of Net PSN, which is reduced by a much increased rate of RESP.

Low light intensities, as noted, will result in a reduction of Total PSN through a reduction in the pool of reduced carriers for the dark reaction. Since under these conditions PSN is light-limited, the overall process is insensitive to temperature changes, there being sufficient heat and CO_2 in the environment. This insensitivity to temperature at low light intensity may be seen in Fig. 10-2, which relates light and temperature to the rate of Total

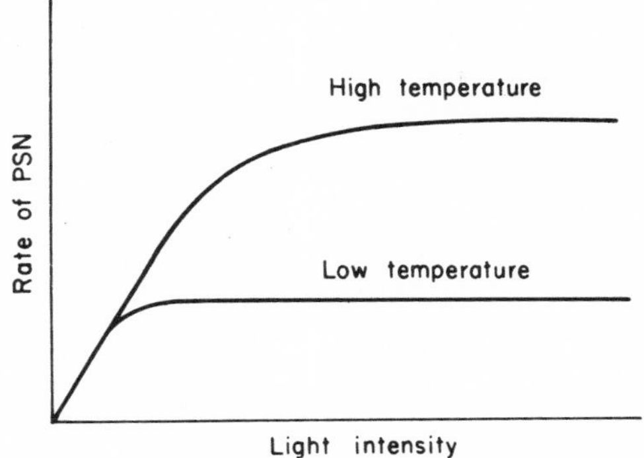

Fig. 10-2. Relationships of the rate of total photosynthesis to leaf temperature and light intensity.

PSN. In the same figure may be seen the sensitivity of the dark reaction, and thus of PSN, to temperature under conditions of higher light intensity. Except as it may cause undue increases in leaf temperature, high light intensities will not result in a reduction in the rate of Total PSN.

When the third limiting environmental factor, CO_2, is in short supply, the rate of the dark reaction is limited. The resulting rate of Total PSN under these CO_2-limited conditions is insensitive to any increases in either light or heat. CO_2 enrichment of the environment, on the other hand, will produce startling increases in the rate of Total PSN under conditions of sufficient light and heat. Prolonged CO_2 enrichment, however, is likely to produce injury to plants.

The effects on the rate of Total PSN caused by insufficient and excessive supplies of heat, light, and CO_2 have been examined. Under what natural conditions might one expect to find these various limitations? Low temperatures in the presence of sufficient light and sufficient CO_2 often occur in alpine environments. Insufficient concentrations of CO_2 in the presence of warm temperatures and high light intensities occur often, for example, within the canopy of a row crop such as corn. Carbon dioxide within the canopy is depleted by high rates of PSN under conditions in which low wind speeds do not replenish the CO_2 by convection. Finally, low light intensities in the presence of abundant heat and CO_2 are found every day in the understory of a forest or woodland. Such conditions are found also in the inner and lower canopies even of dominant plants in a stand.

This matter of different light intensities found at the same time in different parts of a canopy brings up the subject of "light saturation," which is the condition of a leaf or a plant receiving that amount of light or more which will produce the maximum possible rate of PSN under prevailing supplies of CO_2 and heat. Probably for both genetical and anatomical reasons, some plants (called "sun plants") require as much as 30% of full sunlight to become light saturated, while others ("shade plants") require only about 10% of full sun. In general, agronomic crop plants and overstory woodland species are sun plants, while shade plants are found in the understory. Both sun plants and shade plants have about the same "compensation point," which is the light level at which Total PSN and RESP are equal: zero Net PSN. This situation is usually expressed quantitatively in terms of light intensity, which in turn implies a certain rate for the light reaction.

As mentioned in Chapter 8, inner leaves and outer leaves of a plant's canopy have very different energy budgets during any given period of time. So do these inner and outer leaves become light saturated at different light intensities incident on the whole plant. As mentioned, outer leaves may become light saturated in only 30% or so of full sun; but at these intensities the inner leaves may be too shaded even for the light reflected and transmitted to them by the outer leaves to produce light saturation It is only as intensities incident on the whole plant approach full sun that the inner leaves approach light saturation The result of this may be seen in Fig. 10-3a, which shows that, while an individual leaf may be light saturated at moderate light intensities, Total PSN for a whole plant continues to rise as light intensity increases. This in

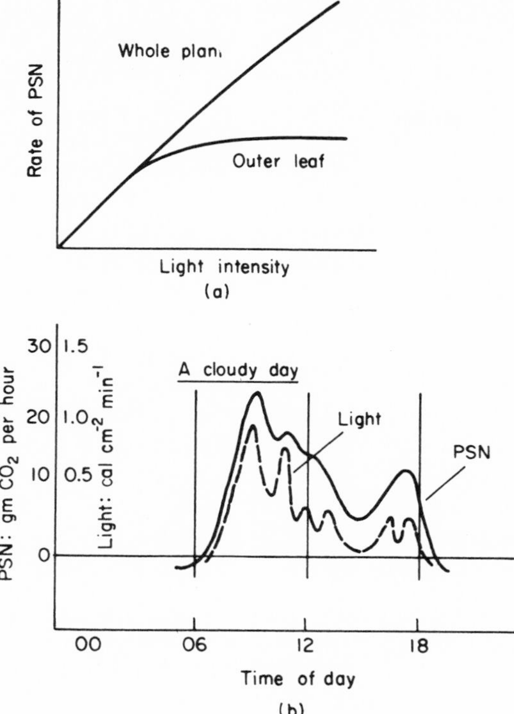

Fig. 10-3. (a) Relationship of light intensity to rate of total photosynthesis in outer leaves compared with the plant as a whole. (b) Relationships between varying light intensity and rates of net photosynthesis on a cloudy day. The units of measurement, PSN in gm of CO_2 per plot per hour and light in cal/cm^2/min, and the values shown, are typical.

turn results in the often-observed rise and fall of the rate of PSN with light intensity through the course of a day, as in Fig. 10-3b.

As noted previously, factors in the soil environment affect the rate of Total PSN, which falls slightly as soil becomes colder or drier. This adverse effect is probably due to increasing viscosity of water and decreasing rates of chemical reaction in the case of cold soils. Dry soil, on the other hand, simply reduces the water available for the light reaction.[1]

Mention was made above of the capture of light energy by chloroplasts of green plants. The substance within the chloroplasts which actually performs this wonderful capture and conversion of light is chlorophyll (CHL). While the absorption spectrum of CHL shows very high absorptivities in some wavelengths, the overall efficiency of PSN relative to the total amount of incident light energy is only about 2 to 5%. CHL absorbs strongly[2] in the blue (about 92% at 0.45 micron) and in the red (about 84% at 0.65 micron), but less strongly in the green (about 70% at 0.55 micron), as mentioned in Chapter 3. As seen in Fig. 10-4, although leaves

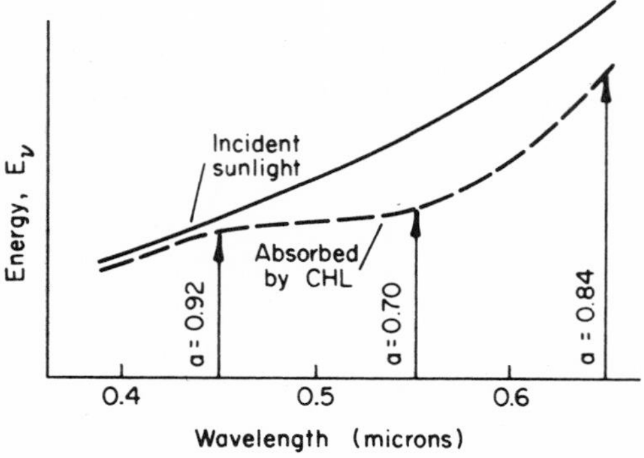

Fig. 10-4. The absorption spectrum of chlorophyll in the region of visible light.[2]

appear green to the eye because of larger amounts of light energy being reflected and transmitted in the portion of the spectrum near 0.55 micron than near 0.45 or 0.65 micron, they may actually be absorbing more energy in the green than in the blue on an absolute basis. Note that the ordinate in Fig. 10-4 refers energy to a frequency, rather than a wavelength, basis, in light of the discussions in Chapter 3. Thus, the curve marked "incident sunlight" continues to rise at wavelengths longer than 0.5 micron. One further note of caution: This figure presents an absorption spectrum for leaf materials rather than an "action spectrum" sometimes encountered in the physiological literature. The former refers to the relative amounts of incident energy absorbed, while the latter refers to the relative conversion of absorbed energy to chemical energy.

Transpiration

The flow of water vapor from the mesophyll through the stomata outward to the atmosphere is called transpiration. While it appears to be principally a necessary by-product of an efficient, gas-exchanging leaf anatomy, this continual loss of water provides portions of the mechanism for heat dissipation and of the mechanism for maintaining liquid water flow through the roots to all plant parts. TRP requirements of plants are far greater than any other water requirement they have. It has been estimated that during its period of growth, an acre of thrifty corn will use 11 acre-inches of water for TRP, exclusive of water lost to evaporation from the field's soil surface.

The rate of TRP may be estimated by keeping track of weight changes in plants or plant parts isolated from a supply of water, and by attributing weight loss to the loss of an equivalent weight of water by TRP. For plants and parts not isolated from a water source, TRP rate may be estimated by gas exchange methods in which concentrations of water vapor are monitored in air entering and in air leaving a controlled chamber containing the transpiring plant or plant part. Though in theory such methods should provide realistic estimates of TRP, in fact the experimental procedures appear to modify the environment surrounding the plant so greatly that TRP estimates thus obtained should be accepted only with caution as representing TRP in the field. Doubtless, most changes in TRP rates so obtained would be qualitatively very similar to those in the field, but the values of the rates themselves are open to question.

Basically, TRP rate is directly proportional to the forces producing movement of liquid and of gaseous water, and inversely proportional to the resistance to water movement in the various parts of the soil-plant-atmosphere system. As in a series system of water pipes, the volume flow rate for the whole system is that of the part with the smallest flow. Factors in the less constricted portions adjust themselves to those in the most constricted, so that the TRP rate must be a series of ratios of driving force and resistance:

$$\text{TRP rate} = \xi_T = \frac{C_1(P_s - P_p)}{R_s + R_p} = \frac{C_2(e_m - e_a)}{R_m + R_a}. \qquad \textbf{(10-6)}$$

Here the first ratio (involving water potential, P) describes the liquid water flow in soil and vascular tissue, while the second ratio describes vapor flow of water outward from the leaves to the air outside the stomata. The driving force for liquid flow is the difference in potential (negative pressure) between that found in the soil, P_s, and that found in the upper portions of the plant near the mesophyll, P_p. Resistances are encountered in the soil, R_s, and in the vascular tissues of root and stem, R_p.

The driving force for vapor flow is expressed as the difference in vapor pressure between that in the mesophyll, e_m, and that in the bulk air, e_a. Resistances are encountered in the mesophyll-stomatal path, R_m, and in the boundary layer of air outside the stomata, R_a. The proportionality constants permit expression of the driving forces in terms of P and e rather than in the less convenient form of water density or concentration.*

To examine environmental effects on TRP by means of Eq. (10-6), consider first the effects on vapor flow. The *TRe* diagram introduced in Chapter 5 gives the form of the relationship of vapor pressure to other environmental variables, and in the case of the saturated air within the mesophyll the relationship of e_m to leaf temperature. What forms do the variations in R_m and R_a take?

In a general way, the resistance presented to vapor flow by the stoma is inversely proportional to the cross-sectional area of the stomatal aperture itself. If the area were circular, the relationship between stomatal diameter and resistance would be $R_m \sim 1/d^2$.

* The units of resistance are those of the ratio (concentration/transpiration rate). Since $\xi = E/L_v$ (Eq. 8-5), in the units used in this book, those of resistance become

$$(R) = (\text{gm/cm}^3)/(\text{gm/cm}^2/\text{min}) = \text{min/cm}.$$

But the aperture is more nearly elliptical near closure, and circular only near full opening. Therefore, the relationship between R_m and the relative degree of stomatal opening is like that shown in Fig. 10-5, where aperture shapes are sketched to show the basis

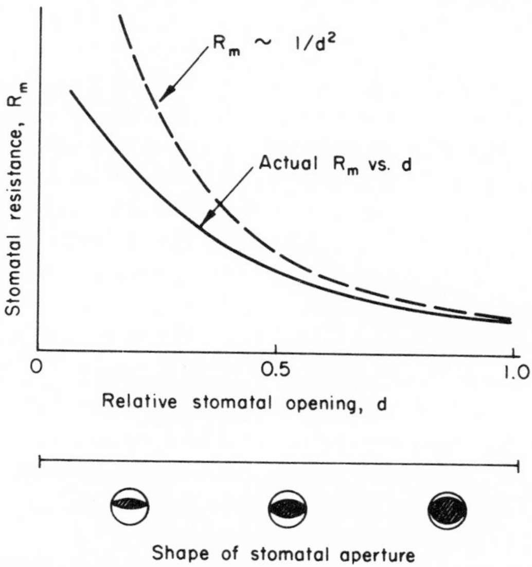

Fig. 10-5. Generalized relationship of stomatal resistance to vapor flow, R_m, and the relative amount of stomatal opening.

for the departure from the curve for $R_m \sim 1/d^2$. When stomata are nearly closed, a small additional opening produces a much larger decrease in R_m, and thus an increase in TRP, than does a comparable opening of already-open stomata.

Experimental evidence is that the resistance offered to vapor flow by the boundary layer of air is approximately of the form

$$R_a = (\text{constant})(D/V)^{1/2}, \quad \textbf{(10-7)}$$

where D is the characteristic leaf dimension and V is the wind speed.[(3)] Thus, the relationship of R_a to V is, qualitatively at least, like that of R_m to the relative stomatal opening, d, in Fig. 10-5. The same experimental evidence shows that, as a general rule, R_m

is about ten times as large as R_a, typical magnitudes for a broad leaf ($D = 10$ cm) being of the order of 0.1 sec/cm for R_a and 1.0 sec/cm for R_m.

With some idea of how the factors involved in control of transpirational vapor flow behave, we may examine the effects of light, wind, moisture, and temperature on TRP by means of Eq. (10-6).

Light affects TRP in two ways. First, light energy falling on a leaf may produce a heat overload under proper conditions. As a result, leaf temperature will rise, e_m will increase, and the vapor pressure difference in the numerator of Eq. (10-6) will increase with a resulting increase in TRP. Second, light falling on the leaf will cause the stomata to open (other conditions permitting) with a resulting increase in TRP as R_m is reduced.

A model explaining the way in which light results in an opening of stomata begins with morning's first light falling on a leaf whose stomata are essentially all closed.(4) With first light, PSN begins and CO_2 within the mesophyll enters the dark reaction. Whether because CO_2 replenishment is impossible through closed stomata or because of increasing concentrations of a by-product of PSN (glycolic acid), the osmotic concentration in the guard cells increases. It rises above that in the neighboring cuticular cells, which have no CHL and thus are not involved in PSN, and a resulting inflow of water produces turgor in the guard cells and stomatal opening. Both PSN and TRP then increase as gas exchange through the stomata is resumed. In the event wilting has produced an insufficiency of water to provide the inflow from neighboring cuticular cells, the stomata do not open. When the rate of water flow through the vascular system is sufficient to allow the inflow to guard cells, the value of R_m for the whole system will be reduced and TRP will resume.

Wind affects TRP in a manner shown in Fig. 10-6. With zero radiant heat loads, increasing wind speed increases convective transfer of heat from warmer air to cooler leaves.(5) This slight transfer results in small increases in leaf temperature, small increases in e_m, and thus small increases in TRP. With high radiant heat loads, the reverse happens: Convective cooling of the leaf,

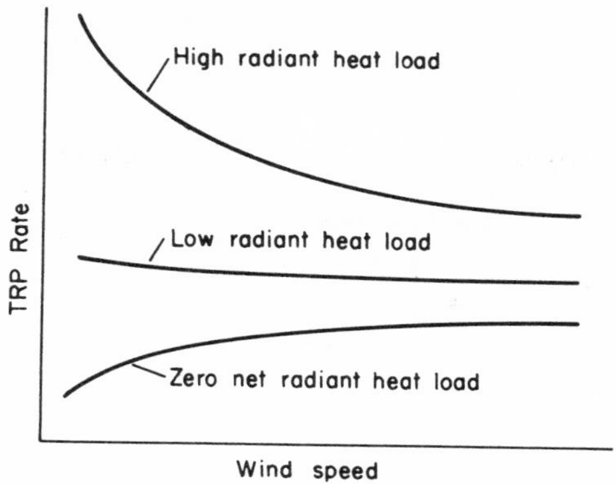

Fig. 10-6. Effects of wind speed on rate of transpiration under different conditions of radiant heat load.[5]

lower leaf temperatures and e_m, and a reduction in TRP. Intermediate heat loads find essentially no response of TRP to wind speed. In all cases, it should be noted, wind will prevent accumulation of water vapor just outside the stomata, which may be viewed in Eq. (10-6) as a reduction in either R_a or e_a. In this sense, wind always acts to increase TRP, as suggested in the discussion of Eq. (10-7). With high radiant heat loads, however, this effect is overshadowed by the effect of wind on leaf temperature, and therefore on e_m.

In summary of the effects of wind on TRP we may note that, under any conditions, wind seems to have its effects in changing TRP only at low speeds. In addition, we should remark that a leaf is almost never entirely free of wind effects, since even in a very calm environment free convection is produced as buoyancy forces move air upward or downward past a leaf, depending upon whether the leaf is warmer or cooler than the environment.

Atmospheric moisture content determines the value of e_a, so that in dry air the lower value of e_a results in an increased value of TRP. As transpiration increases in dry air, however, the increased evaporational cooling of the leaf reduces e_m. Thus, the vapor pressure gradient is not as large as the decrease in e_a by itself might indicate.

Table 10-1
TYPICAL VALUES ILLUSTRATING THE EFFECT OF A GIVEN AIR TEMPERATURE INCREASE ON THE VAPOR PRESSURE DIFFERENCE $(e_m - e_a)$

	Environmental conditions			Leaf conditions			
	T_{air} (°C)	Rel. hum. of air (%)	e_a (mb)	T_l (°C)	e_m (mb)	$(e_m - e_a)$ (mb)	Change in $(e_m - e_a)$ due to 5°C rise in T_{air} (mb)
Night	5	100	8.8	5	8.8	0	
	10	72	8.8	10	12.2	3.4	+3.4
Cloudy day	20	60	14.0	20	23.4	9.4	
	25	44	14.0	25	31.8	17.8	+8.4
Sunny day	20	60	14.0	26	33.6	19.6	
	25	44	14.0	31	44.8	30.8	+11.2

As mentioned previously, light affects TRP partly by way of its effect on leaf temperature. Leaf temperature, and thus e_m, is related to the temperature of the environmental air as well. By means of Table 10-1 and Fig. 5-4, examine the effects on the vapor pressure difference $(e_m - e_a)$ of a 5°C rise in air temperature. Choosing reasonable values for environmental variables for nighttime conditions first, we see that for air and leaf temperatures of 5°C—$(T_1 - T_a) = \Delta T = 0$—and relative humidity of 100%, the vapor pressure difference is zero. If the temperature of the air should increase by 5°C without a change in ambient vapor pressure, the leaf temperature and e_m would also rise and produce an increase in $(e_m - e_a)$ of 3.4 mb.

On a cloudy day with $\Delta T = 0$, as shown in Table 10-1, a rise of 5°C in air temperature from 20 to 25°C, with relative humidity a typical 60% at 20°C, produces an increase of 8.4 mb in the vapor pressure difference of Eq. (10-6). And on a sunny day when leaf temperatures rise 6°C above air temperatures ($\Delta T = 6$°C), the same 5°C increase in air temperature from 20 to 25°C produces an increase of 11.2 mb in the vapor pressure difference. Assuming the resistances in Eq. (10-6) are about the same in all three situations in Table 10-1, increases in TRP resulting from the same temperature rise will be two- and threefold more if they take place during a cloudy and a sunny day as compared with nighttime. Clearly, then, if one analyzes the effect upon TRP of a particular rise in air temperature, he must take care in accounting for the initial air temperature and for whether or not much light is involved.

Soil-Plant Relations

From the vast field of study concerning the plant's interactions with the edaphic portion of its environment, we will be content to examine only a few ideas. First of all, a brief listing will show the complexity of the interactions brought about by the variety of roles played by the soil in the life of the plant: (a) mechanical support, (b) nutrient reservoir, (c) water reservoir, (d) heat reservoir, and (e) source of oxygen-containing air for root respiration.

The slight depression of PSN by cold and dry soil was mentioned briefly in a previous section. The increased viscosity of water and the reduced reaction rates in cold soil also affect TRP, as does the drying of soil. These three processes have the effect of increasing R_s in Eq. (10-6). The increase in the denominator of the portion of the equation describing liquid flow suggests the way in which TRP is decreased.

If the liquid flow becomes the limiting part of the soil-plant-atmosphere system, the factors governing gaseous flow must adjust to make the flow there equal to the liquid flow. Perhaps the major adjustment is the increase in R_m as the stomata close in wilting leaves. Another adjustment is the reduction in heat load on a wilted leaf by virtue of a changed orientation relative to the direct solar beam. A reduced heat load results in a lowered leaf temperature, a lowered value of e_m, and thus a reduced numerator in the vapor portion of Eq. (10-6). Figure 10-7 suggests the net result

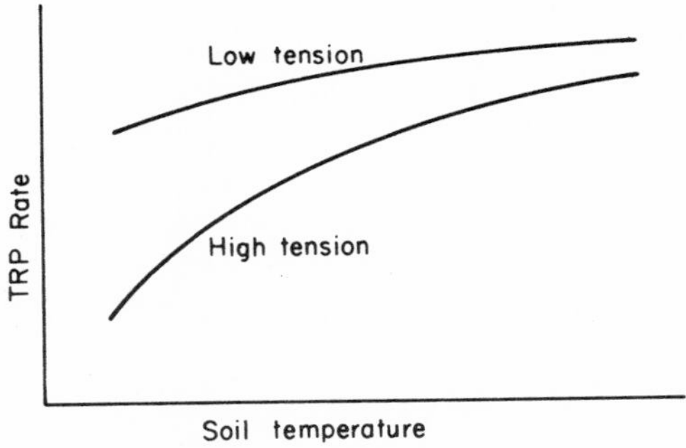

Fig. 10-7. Generalized relationship between soil temperature and tension and rate of transpiration.[6]

of these adjustments due to varying temperatures and moisture contents in the soil.[6] It is quite likely the exact form of the relationships in Fig. 10-7 depends upon such things as plant age, species, and leaf morphology (compare (1) and (6)), but there is no doubt that TRP is reduced in cold, dry soil for the reasons given.

While it is relatively easy to assess the soil temperature in the immediate neighborhood of the water absorbing tissue of roots, it may be rather difficult to estimate the soil moisture tension there.

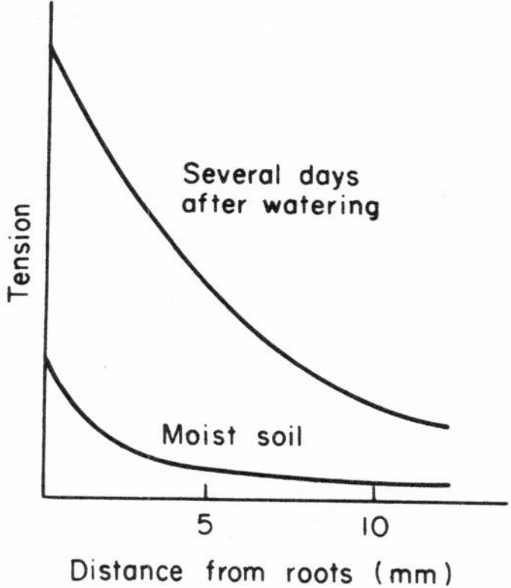

Fig. 10-8. Soil moisture content as a function of distance from roots and time since soil was saturated.[7]

Figure 10-8 shows the results of some carefully conducted laboratory research[7] which suggest that measurements of tension made away from the immediate root mass of a plant may greatly underestimate the tension experienced by the roots themselves. As would be expected, roots extract the water from their immediate soil environment. If conditions prevent rapid enough root tip growth into regions of moist soil, gradients of tension develop which may not be detected with many methods of measuring soil moisture.

Finally, we wish to note the way in which soil salinity acts to determine the soil "texture." As seen in Fig. 10-9a, the addition of salts changes the soil moisture-tension relationships in such a way as to make the soil more clay-like. For a given water ratio, addition of salts produces an increase in the tension by virtue of increased osmotic forces opposing extraction of soil moisture by roots. Figure 10-9b depicts experimental results,[8] which demonstrate the fact that the forces described in Chapter 5 are additive in producing the resultant force opposing the roots' extraction of soil moisture. In the experiments, a variety of soil moisture and salinity

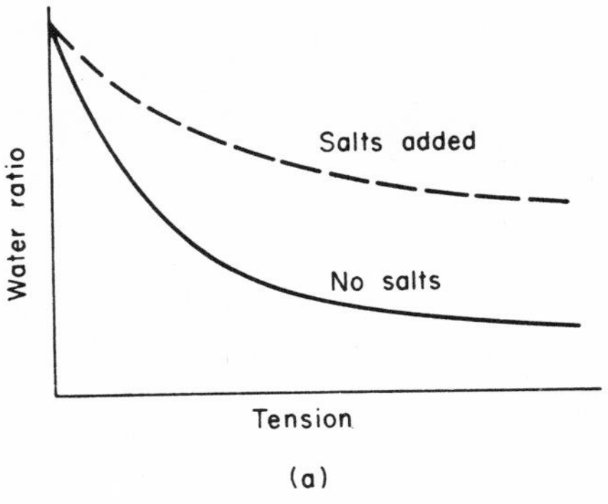

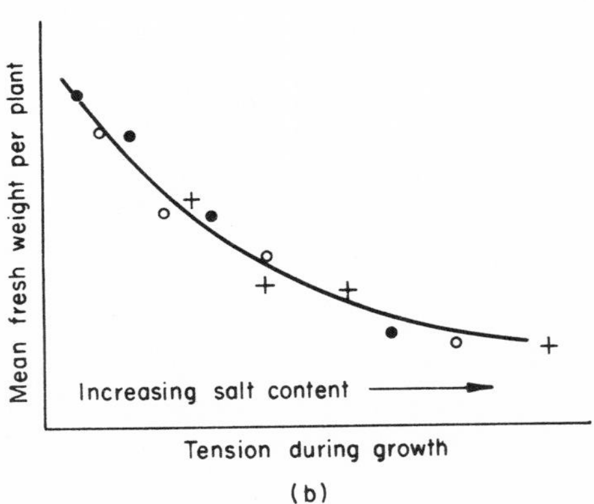

Fig. 10-9. (a) Effect of added salts on the moisture-tension curve of a typical soil (b) Experimental evidence that the components of tension are simply additive so far as plant growth response are concerned.[8] Key: ●—wet soil, O—moist soil, and +—dry soil.

conditions was used to produce soils having a range of effective tensions. Seedlings were grown with constant soil moisture in each of the soils for a period of time, and their growth was expressed as the mean fresh weight per plant for each soil condition. The results show that growth had the same inverse relationship to tension regardless of the combination of moisture and salinity which produced the tension.

REFERENCES

1. O. Babalola, M.S. Thesis, Department of Soils, Oregon State Univ., Corvallis, Oregon, 1967.
2. See e.g., D. M. Gates, *Appl. Opt.* **4**, 11 (1965); and W. E. Loomis, *Ecology* **46**, 14, (1965).
3. R. O., Slatyer, *Agr. Meteorol.* **3**, 281–292 (1966).
4. See e.g., P. E. Waggoner and I. Zelitch, *Science* **10**, 1413 (1965).
5. K. Knoerr, *Proc. Ann. Tech. Meeting, Inst. Environ. Sci.*, pp. 615–623 (1965).
6. L. M. Cox, and L. Boersma, *Plant Physiol.* **42** (4), 550–556 (1967).
7. L. M. Cox, Ph.D. Thesis, Department of Soils, Oregon State Univ. Corvallis, Oregon, 1966.
8. C. Wadleigh, *Yearbook Agr.* p. 358 (1955).

PROBLEMS

10-1. If warm air is advected to the neighborhood of a sunlit leaf (i.e., heating not due to increased radiation load) with no change in moisture content of the air, explain the changes you would expect in the rates of transpiration and photosynthesis in the leaf.

10-2. Verify the entries of Table 10-1 graphically by plotting them on a *TRe* diagram.

10-3. If the surface in Problem 8-1 dissipates by evaporation 0.20 cal cm^{-2} min^{-1} with a 5°C temperature difference in air with a relative humdity of 50%, plot the following three variables as functions of increasing surface-air temperature difference;
(a) Energy dissipated as E_{uo} (cal cm^{-2} min^{-1}),
(b) the Bowen ratio, and
(c) the total heat load assuming $B = M = 0$.

10-4. Evaluate the constant of proportionality in Eq. (10-7) from the "typical magnitudes" presented in that paragraph. On axes of leaf dimension and wind speed, construct isolines of R_a for values of 0.1 sec/cm and 0.5 sec/cm.

10-5. You are given the following observations:
(a) In a weather shelter the temperature is 60°F and the relative humidity 60%.
(b) The temperature of a nearby sunlit leaf is 65°F.
(c) A nearby shaded leaf is transpiring at a rate half as great as the sunlit leaf.
(d) The shaded leaf has a total resistance to vapor flow which is 50% greater than that of the sunlit leaf.

What is the temperature of the shaded leaf?

Chapter 11

PLANTS AND THE ATMOSPHERE: GROWTH AND DEVELOPMENT

Introduction

IN THIS CHAPTER, our attention turns to the net results over longer periods of time of the kinds of hour-to-hour and day-to-day variations in physiological processes discussed in Chapter 10. To begin, we should make clear the distinction between the terms "growth" and "development." Growth refers to the accumulation of mass or the increase in size of a plant through time. The mass may be either "fresh" (plant materials + plant water) or dry (plant materials alone); the size may be variously overall height, total leaf area, total length of root system, etc. Development refers to the passage of a plant through a sequence of morphological and functional stages, regardless of whether its growth is great or small. Changes from one stage to another in the sequence are marked by such phenomena as bud-burst, flowering, fruit or seed set, etc.

Temperature requirements for best growth and development are usually expressed in terms of optima and limits. Optima represent the best possible conditions of temperature, while limits represent the extremes of possibility for maintenance of life. It is not entirely clear whether temperature by itself or temperature as an integrated measure of both heat and light exerts its control over plant growth and development. It is clear, however, that the temperatures defining limits and optima change with time as the plant ages and passes through the sequence of developmental stages. In fact, at any one time, limits and optima for one process (e.g., leaf elongation) may be different from those for another (e.g., differentiation of flower primordia).

Temperature limits and optima may be dependent upon the past history of the plant's environment—so-called "pre-conditioning." For example, under proper conditions low temperatures may cause severe physical damage to a plant by the rupturing of individual cells when ice crystals form within them. On the other hand, if the same plant had experienced a proper number, sequence, or frequency of moderately cold periods prior to the low temperatures in question, the damage might never occur because of a process known as "frost-hardening." Incidentally, day length may be a primary environmental control on frost-hardening in some species, but the general point about preconditioning remains valid.

In addition to various temperature effects on plant growth and development, there are effects which are related variously to the intensity, the quality, and the timing of light and water. In this chapter, we will mention only a few of the patterns of growth and development resulting from the myriad possible combinations of key variables of the physical environment. To these key variables we would have to add, in any complete discussion, such variables of the biological environment as competition for heat, light, water, and nutrients by other plants and animals; the processes of symbiosis and parasitism; and even suspected inhibitions of one species by exudations from other neighboring species. In brief, the list of interactions of a plant with its physical and biological environment is almost endless.

Thermoperiodism

Plant growth is frequently enhanced in an appropriately fluctuating temperature environment as compared with any single, constant temperature. To this fact, the term "thermoperiodism" has been applied. When the fluctuations in temperature are between night and day, we speak of diurnal thermoperiodism.

What evidence do we have that diurnal thermoperiodism exists, and what might be the cause? In Fig. 11-1 appears a drawing, similar to the famous one by Went,[1] presenting the results of a hypothetical series of experiments carried out under controlled greenhouse conditions. Each data point plotted shows the total

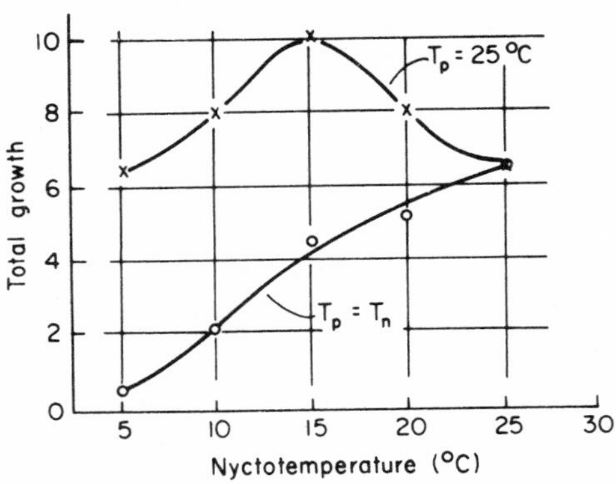

Fig. 11-1. Typical data providing classical evidence of thermoperiodism.

growth during the period of the experiment for a plant or group of plants which grew under the controlled conditions of dark-period nyctotemperature, T_n, and light-period phototemperature, T_p, indicated. One set of plants grew at constant temperatures, while the other set grew at the same phototemperature and different nyctotemperatures. Several of the experiments in fluctuating temperatures produced growth which was better than any produced in a constant-temperature environment, and one experimental combination of temperatures clearly produced the best growth. Went's figure and others similar to it embody the substantive evidence for the existence of diurnal thermoperiodism.

What of the explanation for thermoperiodism? In Fig. 11-2a appears an often-observed relationship, first reported by the famous plant physiologist Sachs,[2] between growth rate of a species and the temperature at which it grew. There is in the figure a single intermediate temperature which is optimal, and the hypothetical explanation is that there are two growth-limiting, temperature-dependent processes whose combined effect yields the curve shown.

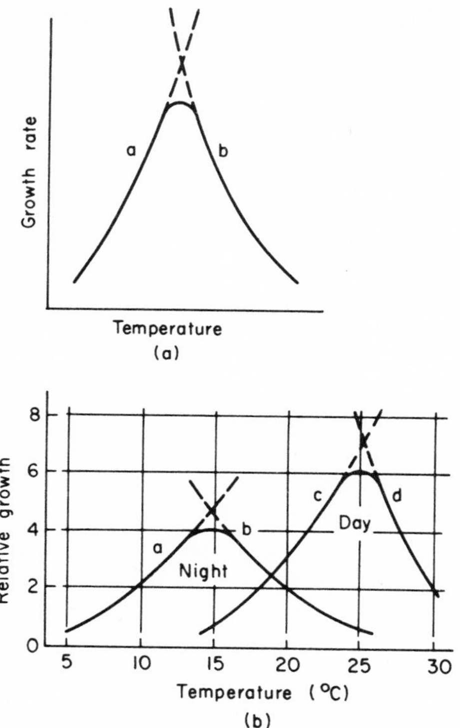

Fig. 11-2. (a) Relationship between growth rate and average temperature during the growing period. (b) Hypothetical relationships between growth rate and mean temperature during nocturnal and diurnal portions of the growing period.

The temperature relationships of these two processes are indicated by the curves marked *a* and *b*, and we see that the benefits of increasing temperature in process *a* are overshadowed at the overall optimum temperature by the deleterious effects of increasing temperature in limiting process *b*. While the rates of growth and the actual optimum temperatures vary with species, this general relationship has been observed for a great variety of plants.

In view of the often distinct differences in energy budgets and physiological processes taking place in plants during the day as compared with the night, it seems reasonable there might be a basis for hypothesizing relationships such as those shown in Fig. 11-2b. Here, instead of the combined growth rate and average temperature of Fig. 11-2a, daytime growth rate is compared with phototemperature and nighttime growth rate with nyctotemperature. Careful examination will show that the growth rates shown for various combinations of T_p and T_n in Fig. 11-2b are exactly those of Fig. 11-1. For example, in an environment characterized by temperatures changing diurnally between $T_p = 25°C$ and $T_n = 20°C$, the relative growth rates would be 6 by day, 2 by night, and their sum of 8 that shown in Fig. 11-1 for the particular temperature combination.

The hypothesis represented by Fig. 11-2b is only as good as the likelihood the limiting processes *a*, *b*, *c*, and *d* are real. The fact is, we do know of processes which might well be the four limiting processes. For example, process *a* might be the one governing nighttime cell elongation, which is known to increase with increasing T_n. Process *b* might represent the result of increased RESP with higher T_n and in turn a lower Net PSN for a 24-hour period. Process *c* might represent either the same process as *a* under conditions with light, or a different process such as the increase in PSN with increasing light intensity (Fig. 10-3), the increasing temperatures here actually reflecting an increase in light intensity. Both decrease in Net PSN with increased RESP, and at higher temperatures, protein denaturation are known to produce results such as those suggested by process *d*; so we are not at a loss to find combinations of temperature-sensitive physiological processes which might yield relationships such as those in Fig. 11-2b. Such a combination, if it indeed exists, would produce evidence of thermoperiodism such as that in Fig. 11-1. If the actual mechanistic explanation of thermoperiodism is not exactly as suggested here, it may well involve the concepts of limiting and overshadowing process interactions discussed.

In the introduction to this chapter, it was mentioned that different species and different aged plants of the same species show different thermoperiodic optima. To begin a consideration of this point, consider the growth data for tomato taken from Went[3] and plotted on coordinates of T_p and T_n in Fig. 11-3. In this "contour" method of presenting data on thermoperiodic response, values of some measure of plant response are indicated by lines of equal response passing through various temperature combinations.

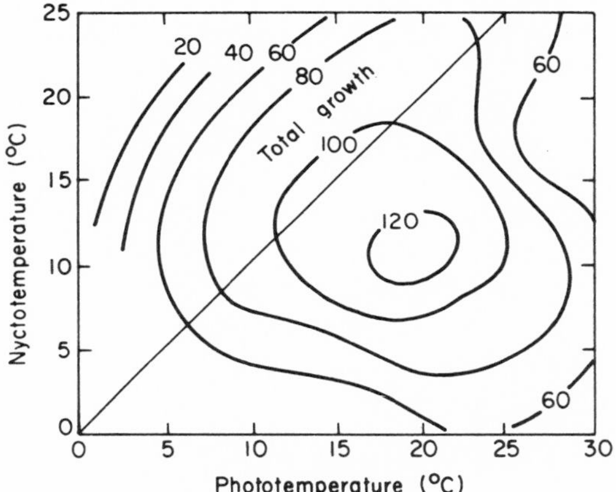

Fig. 11-3. Total growth of tomato seedlings as a function of phototemperature and nyctotemperature. Data are from Went.[2]

The height and location of the "topographic summit" gives the results and temperature associated with the optimal conditions for growth. If the optimum lies off the 45° line representing a constant-temperature environment, the plant represented is, by definition, thermoperiodic.

That different species have different optima and yield different contour orientations may be seen in comparing the three parts of Fig. 11-4[3] with one another and with Fig. 11-3. Plotting experimental data in this manner not only shows the nature of the plant's response to temperature in a rather easily comprehended manner, it also shows at once the additional ranges of temperature most likely to yield valuable information in the future. For example, extensions of experimental temperature into the ranges $T_p = 30°C$ and T_n from 5 to 25°C appear from Fig. 11-4b very likely to disclose a thermoperiodic optimum for Sequoia seedlings of the age studied by Hellmers and Sundahl.[4]

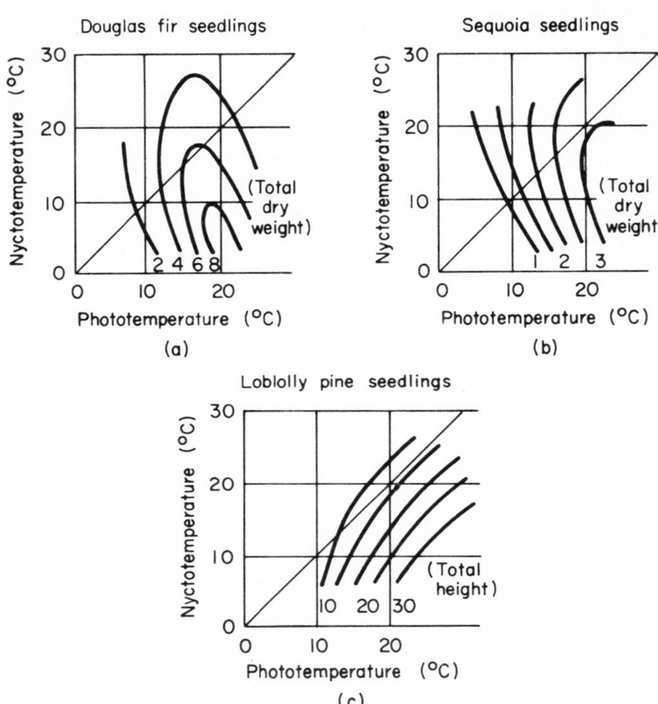

Fig. 11-4. Growth response of three species of coniferous seedling to phototemperature and nyctotemperature. Data on Douglas fir and Sequoia from Hellmers, and on pine from Kramer.(3)

The differences in optima and in contour orientation between plants of different age in the same species are suggested in Fig. 11-5 based on data from Went(5) for tomato seedlings, ages 1 week and 3 weeks. As is found to be the case for a number of species studied, the optimum temperature changes to lower nighttime values with increasing age.

Finally, it should be noted that the location of the optima and the orientation of the contours depend upon the physical conditions, such as light intensity and photoperiod, of the experiment and upon the growth measures used. Hellmers(6) used total top

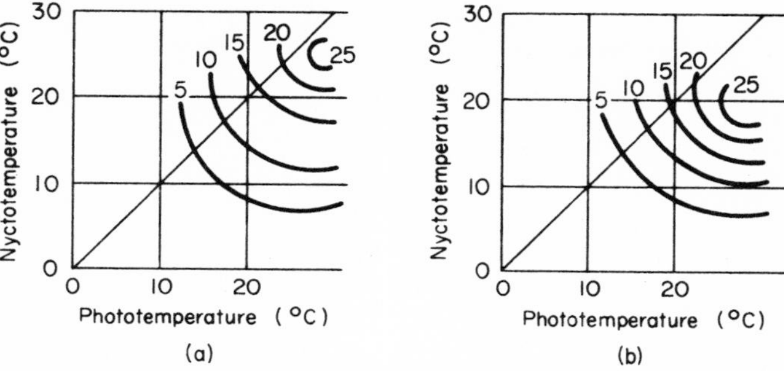

Fig. 11-5. Thermoperiodic growth response of tomato seedlings as a function of age. (a) Age 1 week. (b) Age 3 weeks. Data from Went.[5]

height and total dry weight after a certain period of growth, with the results shown in Fig. 11-6 plotted by the contour method. The contour orientations are distinctly different from those for the same growth measures in other species in Fig. 11-4. In fact, neither growth measure in Fig. 11-6 even suggests there is an optimum or

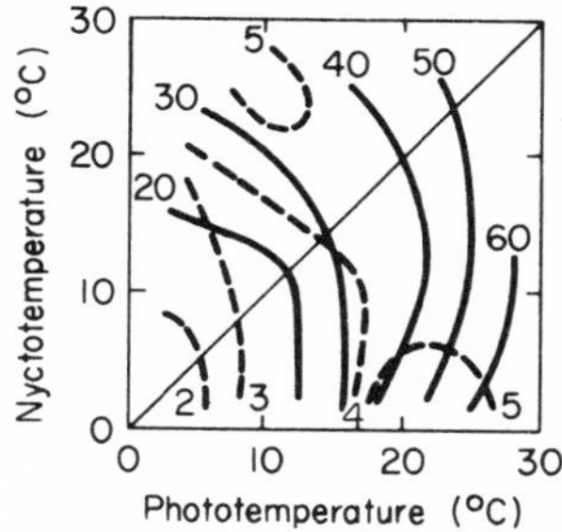

Fig. 11-6. Response of two measures of growth in Jeffrey pine to phototemperature and nyctotemperature. Solid curve—total top height, dashed curve—total dry weight. Data from Hellmers.[6]

that the Jeffrey pine is thermoperiodic. Of equal interest, perhaps, is the large area for temperature ranges of about $15 < T_p < 30$; $10 < T_n < 25$ where Jeffrey pine seedlings appear to put on height growth with no increase in dryweight—a very "leggy" seedling indeed.

Phenology and Tolerance

Phenology is the study of the time patterns associated with the developmental sequences of organisms. Scientists investigating the environmental requirements for development often examine the organism's tolerances to various factors during individual phenological, or developmental, "subperiods," such as that from the emergence of a plant seedling until it begins to flower. As noted in the introduction to this chapter, the tolerances—expressed as limits and optima—are likely to change markedly with age of the organism. We have seen in Fig. 11-5 how the thermoperiodic optimum for the growth rate of tomato changes in the first few weeks after emergence.

Figure 11-7 from Wang[7] shows how the thermal tolerances

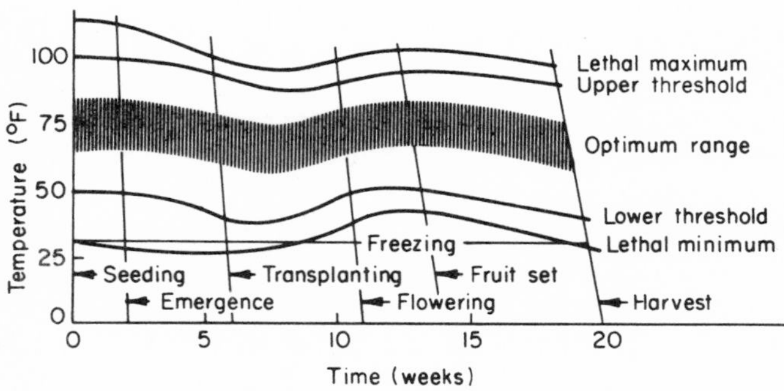

Fig. 11-7. Thermal limits and optima for tomato as a function of development through harvest. After Wang.[7]

of the tomato change through the phenological subperiods from seeding to harvest. Only the curves for "day temperature" have been taken from the original figure, which includes also curves for "night temperature" and "soil temperature." As one would expect, both lie below those for day temperature, with soil temperature between day and night temperatures.

Many things may be seen at once from Fig. 11-7. First of all, limits and optima rise markedly at the time of flowering, having fallen just as markedly as the plant, at about 6 weeks after seeding, grew out of the seedling stage. We can see here the same fall in optima as in Fig. 11-5 if account is taken of the fact that there the ages of 1 and 3 weeks were reckoned from emergence rather than from seeding. Then, too, in Fig. 11-7 we see that, during fruiting, even the lethal minimum temperature rises above the near-freezing values it has at other times in the life cycle. Finally, the slightly slanted vertical lines indicate that less time is required at higher temperatures to pass through a phenological subperiod than at lower temperatures. Wang, in preparing his curves, used experimental data from both field and laboratory obtained by many workers. He has prepared the same kinds of response curves for other plants,[8] both orchard and field crops, but each species has its own configurations slightly different from the others, as in the case of contours for diurnal changes of temperature. There may also be significant varietal differences of response within species.

In addition to the thermal responses of species through the period of growth and development, we have already noted the responses to other environmental factors. Responses to environmental moisture are perhaps the next most-studied after temperature, and because observations of precipitation amount are the most readily available measures of environmental moisture, it is with rainfall that plant responses are often correlated. Azzi[9] has given us a conceptual approach to sorting out the effects of various environmental variables. The scheme for determination of what he calls the "Meteorological Equivalent" provides the basis for examination of this approach. Figure 11-8 depicts Azzi's scheme

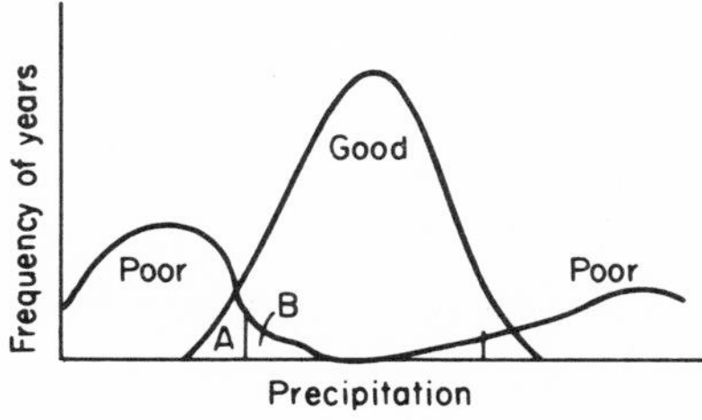

Fig. 11-8. Determination of the meteorological equivalent.

(visually different from his presentation, however) in which are plotted the frequencies of years having very good biological response and very poor biological response against the values for some environmental variable in the individual years. In order to obtain more contrast between "good" and "poor" responses, it is recommended that years having intermediate responses not be considered in the scheme.

How do we interpret Fig. 11-8? We see that poor responses, represented for example by some measure of yield, are associated with both low and high values of the environmental variable, for example, total precipitation. Good responses—yields—are associated with intermediate values of precipitation. Our objective is to make a rational selection of two values of the environmental variable which set the good yields apart from the poor. The smaller value of precipitation is the one which divides the area of overlap into the two equal areas, A and B. Note, this value is not necessarily the one where the "good" and the "poor" curves cross. In this case, the smaller value is called the "Meteorological Equivalent (ME) for precipitation deficit." The larger precipitation value is determined in like manner and is called the "ME for precipitation excess."

Since Azzi draws a figure like Fig. 11-8 for each phenological subperiod and each species or variety, these two additional factors are required for a complete specification of each ME. If Fig. 11-8 represents, for example, the yield of corn as related to the precipitation during the subperiod from tasseling to silking, the complete specification of the ME requires five kinds of information:

(1) the environmental variable (precipitation),
(2) its intensity (deficit or excess),
(3) the species or variety (sweet corn),
(4) the phenological subperiod (tasseling to silking), and
(5) the response measure (yield).

For each combination of these five factors, there will be a separate meteorological equivalent. For any given species, environmental variable, and response measure, the ME's based on the other two factors may be put together to form response curves such as those for the response of tomato yield to day temperature in Fig. 11-7.

The "Heat Units" Concept

Figure 11-2a, although referred to above in connection with diurnal temperature changes, suggests there is a relationship between development rate and temperature when viewed from the standpoint of a whole subperiod or a whole season. Indeed, this often appears as an inverse relationship between temperature and the length of time required to pass through some developmental period, as in Fig. 11-9. The curve tells us there is some low temperature below which the plant will not develop at all and will take an infinite time to pass through the period. Also, there is some intermediate optimum temperature which permits the most rapid development rate. Above the optimum, adverse temperature effects produce a lengthening of the time for development.

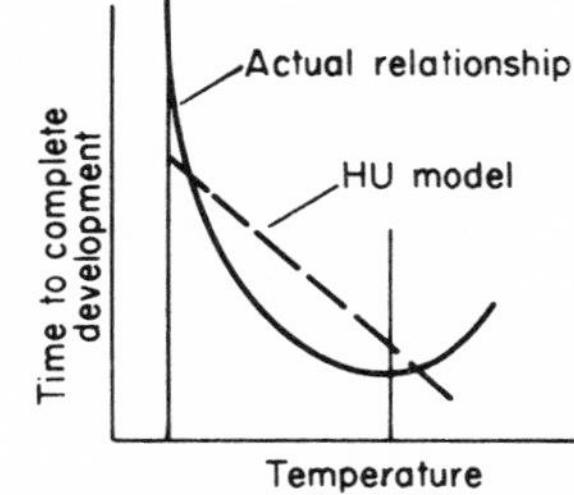

Fig. 11-9. Relationship between development time and mean temperature for the period, as compared with the heat units model for the relationship.

Figure 11-9 must be considered as a simplification of the net results of many temperature-dependent physiological and biochemical processes through an entire subperiod or season of development. It is as if the net results of total PSN, RESP, TRP, and many other, even more complex processes and reactions could be regarded as the results of some temperature-sensitive biochemical "master reaction," whose rate rises as temperature rises. The dangers inherent in simplifying things to this extent are obvious, but for purposes of approximating seasonal temperature effects on plant growth and development, this simplification has been useful and has been employed in a class of procedures known collectively as the "heat units" (HU) concept.

The heart of the HU concept lies in Eqs. (11-1):

$$\mathrm{HU} = \sum_{N} \left[\frac{T_{\mathrm{M}} + T_{\mathrm{m}}}{2} - T_{\mathrm{t}} \right] \qquad \text{when} \quad \frac{T_{\mathrm{M}} + T_{\mathrm{m}}}{2} > T_{\mathrm{t}} \quad \textbf{(11-1a)}$$

or

$$\mathrm{HU} = \sum_{n} (T - T_{\mathrm{t}}) \qquad \text{when} \quad T > T_{\mathrm{t}}. \qquad \textbf{(11-1b)}$$

Here HU is the number of heat units accumulated during either N days, in the case of Eq. (11-1a), or n hours, in the case of Eq. (11-1b). The first equation is employed with daily temperature maxima, T_{M}, and minima, T_{m}, and some threshold temperature, T_{t}. The second equation is employed with hourly values of temperature, T. In the former, results are expressed as "growing degree-days" and in the latter as "growing degree-hours." In some applications of the HU concept, the reversal of the curve at high temperatures in Fig. 11-9 is accounted for by requiring that in Eq. (11-1b) T must also be less than some "ceiling temperature" threshold. The contention of the concept is that the number of heat units accumulated during any phenological subperiod, and therefore also for an entire developmental season, is constant from year to year for any plant variety, or from place to place within any year for that variety. In a cool year, it will take a greater period of time to accumulate the required number of units, and in a warm year, a shorter period. Applications of this concept have employed various values of T_{t} for various plants and have found different values of HU—known as "varietal constants"—for different plants.

Figures 11-10a and 11-10b show graphically the implications of the HU concept. In 11-10a, the two temperature sequences are plotted against time. These two sequences may be viewed as the progress of daily mean temperature from Time 1 to Time 2, either in two different climates or in two years at the same location. Figure 11-10b shows the manner in which heat units accumulate between 1 and 2 according to Eq. (11-1a), for example. The heat units concept says that a plant variety which passes through a certain sequence of development between 1 and 2 in climate or year A will pass through the same sequence in the same time in

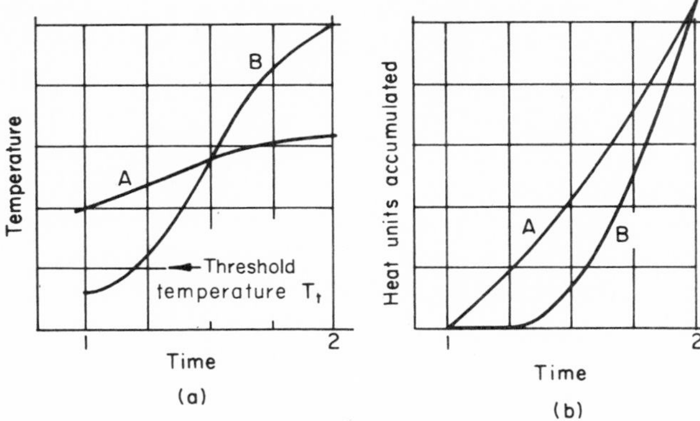

Fig. 11-10. (a) Hypothetical seasonal increases in temperature during two growing seasons. (b) Temperatures of growing seasons expressed as accumulating heat units.

climate or year B, since in both cases the same number of units accumulates between 1 and 2.

We have hinted already that the HU concept may be open to valid criticism. What are some of the details of such criticism? First of all, what happens to the concept when we have a plant whose production is best when it grows in warm spring weather during its early stages and in a cool summer during its later stages each year? In spite of the equality of HU between curves A and B in Figs. 11-10, we would have to conclude the plant would produce better under the conditions of climate or year A. The nature of the thermal responses shown in Fig. 11-7, then, is a basis for criticism of the concept.

It should be noted that the matters of growth and development, and thus of response measure, should be separated in approaching the evaluation of the HU concept. That is, while curves A and B in Figs 11-10 may both produce the same growth between times 1 and 2, they may produce entirely different results in terms of development. An example from the response of corn to moisture, while not referring to heat units, will serve to illustrate the principle in point. Azzi[(9)] has found that corn is quite sensitive to the amount of precipitation during the period of about two weeks preceding earing. If the precipitation is in the optimal range between the ME's for deficit and excess during this period, the production of viable seed will be roughly proportional to the amount of precipitation during the period following earing. If the precipitation during the critical period is suboptimal, however, plentiful rainfall afterward may produce ultimate growth as lush as in the first case, but there will be no seed production. Thus, in evaluation of seasonal effects of environmental variables as in the HU concept, one must separate growth from development. In fact, in really careful evaluation, one would have to separate development expressed as, for example, yield of seed from development expressed in terms of the quality of the seed.

A second criticism of the HU concept lies in recognition of the nonlinearity of the curve in Fig. 11.9. Even if the use of Eqs. (11-1) employs an upper threshold temperature (which implies an infinite development time in hot environment as well as cold), the fact that the HU concept assumes implicitly there is a linear relationship between temperature and plant response leads to trouble. Accumulation of more heat units on a given day than on the day before may or may not mean enhancement of growth on the second day. It depends, among other things, upon whether the mean temperature the first day was at or below the optimum.*

A third criticism is that net responses of plant growth and development are to the temperatures of the plant parts themselves, and these may be quite different from temperatures measured in a weather shelter some distance from the plant. As seen in Chapter 8, the temperature difference between plant and air may be quite variable at a given time, and even though small, the cumulative effects through an entire growing season may be quite large.

Applications

Most applications of the heat units concept fall into the broad category of forecasts of crop harvest dates, yields, or quality. Wang[8] mentions both the variety of crops and the numerous modifications of the basic concept which have been reported on.

* The dashed line in Fig. 11-9 represents the assumption made by the HU concept.

He points out also that the concept has been applied to other than crop plants[10] and to the problems of growth and development in insects, plant pathogens, birds, and other animals.

A potential area of application often mentioned, but seldom exploited so far as the writer knows, is the use of the HU concept in estimating the likelihood of successful growth of a crop in an area for which weather data are available but in which the crop has not been grown before. Closely related to this application are two other, relatively unexploited applications. One is the selection of the optimal planting data for a given area—optimal in the sense that it will, in most years under the climate in question, put the developing plant "in step" with the sequences of weather to follow. The second is the use of climatic data in selecting from among several varieties of a plant to be grown in a new area.

Perhaps the most famous application of the HU concept had success assured in spite of the shortcomings discussed. Thornthwaite[11] noted the confusion and losses resulting from the arrival of canning peas at the cannery all in the course of one or two days. Noting also that the fields producing these peas were also all planted during the course of a few days, he suggested planting dates be spread over a month in order to produce a more orderly arrival rate, spread over a week or so, at the cannery. Because the HU concept really is valid up to a point, and because a more rapid accumulation of units near the warm end of a growing season than at the cool beginning produced the desired schedule of maturation and harvest, the application was a success. It was worth so much to produce an effective use of cannery facilities by producing an orderly flow of peas, the fact that the HU concept did not produce an accurate forecast of harvest date in any given year was almost immaterial.

Another potential area of application of the ideas presented in this chapter is the modification of the microenvironment of individual organisms in such a way as to produce nearly optimal conditions at each point in the developmental cycle of the organism. In the next chapter are discussed several of the kinds of modification employed already, but many feel these represent only a beginning in the biometeorological management of organisms of all kinds.

REFERENCES

1. F. W. Went, *Amer. J. Botany* **31**, 135 (1944).
2. J. Bonner and A. W. Galston, "Principles of Plant Physiology," p. 460, Freeman, San Francisco, California, 1952.
3. F. W. Went, *Amer. Scientist* **44**, 378 (1956).
4. The data for Douglas fir and for Sequoia are from H. Hellmers and W. Sundahl, *Nature* **184**, 1247 (1959), and those for Loblolly pine from P. Kramer, *Forest Sci.* **3**, 45 (1957).
5. F. W. Went, *Amer. J. Botany* **32**, 469 (1945).
6. H. Hellmers, *Forest Sci.* **9**, 189 (1963).
7. J. Y. Wang, "Agricultural Meteorology," p. 350, 1967. This book contains exhaustive and informative discussions of research on the various climatological effects on growth and development. At this writing the book is available from Dr. J. Y. Wang, San Jose State College, San Jose, California.
8. Thermal responses of corn may be found on p. 352 of Wang's book,[7] and those for canning peas in *Ecology* **41**, 787 (1960).
9. G. Azzi, "*Agricultural Ecology*," Constable Press, London, 1956.
10. The HU concept, with all its shortcomings, has been applied to observations, for example, of the famous Japanese cherry trees in Washington, D.C., by A. A. Lindsey, *Ecology* **44**, 149 (1963).
11. The late C. W. Thornthwaite was mentioned in Chapter 8 in connection with a method of estimating E for a Type 1 energy-budget system. An almost humorous description of the cannery problem and its solution appears in *J. Operations Research Soc. Amer.* **1**, 33 (1953).

PROBLEMS

11-1. The following are data on growth experiments made under controlled laboratory conditions. Each entry is the mean dry weight per plant for a group of plants grown under the same conditions. T_p and T_n were alternated in periods of 16 and 8 hours respectively.

	T_p (°C)				
T_n (°C)	20	25	30	35	40
5	2	8	40	8	Death
10	15	21	53	40	2
15	29	35	67	54	4
20	15	24	55	38	Death
25	3	9	41	28	Death

(a) Is the plant thermoperiodic with respect to growth?
(b) Estimate relationships of relative growth to T_p and T_n as in Fig. 11-2b.
(c) Plot the growth response of this plant as a function of the mean temperature per 24-hour period. Is the relationship dependent upon the value of $(T_p - T_n)$?
(d) Would you recommend that further experiments be conducted at different temperatures than those shown? Why?

11-2. Following are temperature records from a weather shelter located in an orchard during two growing seasons. In year *A* the record shows an accumulation of 150 degree-days above a threshold of 40°F when fruit matured on Day 10.
(a) On what day were 150 degree-days reached in Year B?
(b) Assuming departures of predicted and actual maturity date due primarily to plant parts being above air temperature, in which year would you say the predicted date and the actual date would agree best?

Day No.	1	2	3	4	5	6	7	8	9	10
A: T_{max}	65	60	80	65	65	60	70	55	85	65
T_{min}	55	40	40	35	55	40	50	45	35	35
B: T_{max}	80	85	80	65	65	70	85	85	80	65
T_{min}	40	35	40	55	55	50	35	35	40	55

11-3. Following are data on the yield of two varieties of a crop plant as a function of precipitation amount during the plant's critical subphase, and on the climate of the locality where the crop is to be grown commercially. Calculate the long-term average yield for each variety and then select the better one for planting. Assume temperatures in this climate are equable during the critical subphase regardless of precipitation amount.

Precipitation during 15 days before flowering is	Variety A	Variety B	Climatological probability at planting site
Below the meteorological equivalent for drought	10	5	0.4
In the "optimum" range	40	60	0.2
Above the meteorological equivalent for excess	30	20	0.4

Chapter 12

ARTIFICIAL CONTROL OF PLANT ENVIRONMENTS

Introduction

AS SUGGESTED at the end of the last chapter, artificial controls placed on the principal variables in an organism's environment could, if properly applied, keep the environment optimal with respect to whatever organismic response is considered important. Clearly, the combination of these variables which would constitute the optimum changes as the organism grows and develops, and, as we have mentioned, according to whether it is growth or development which is of final concern in the management of the system.

In seeking optimal conditions for one organism, the most effective means may be to provide an environment which is harmful to another, neighboring organism. Thus, the artificial controls of which we speak may be intended either to be beneficial to one group of organisms or harmful to another group—and sometimes both of these at once.

Another aspect of artificial controls arises from a consideration of the variation in optima with the stage of development. What is a beneficial control during the period from planting to emergence, let us say, may be quite harmful—or at least quite suboptimal—following emergence. Controls must be considered, therefore, as either "continuous" or "intermittent." That is, they will consist of installations or applications which are permanent or semipermanent in the organism's environment, or which are present only during certain parts of the developmental cycle or during some emergency. As may easily be imagined, it is a fair rule of thumb that the more intermittent the control, the more expensive is its installation and application. At the same time, such controls usually represent a very high ultimate return on the initial investment if the control is successful.

The reader, upon considering the complex interactions within the energy budget and within the physiology of growth and development, can readily understand the complexity of the interactions of environment and organism together. It should be clear at this point that control over one component of the environment may well result in unexpected changes in other components. If harmful, they would be even more harmful because they are unexpected. It is possible, of course, to have the unexpected effects appear directly within the organism of concern rather than within the organism's environment. All successful practices of environmental control require knowledge—as complete knowledge as is possible—of both the organism's complex requirements at a given time and of the modes of physical interaction within and between organism and environment. Rephrasing the concluding remarks of Chapter 8, one may approach through the energy budget the problem of meeting an organism's requirements without producing unexpected and undesirable side effects.

Finally, in this chapter we shall be considering various forms of control presently used in plant environments, and examining their effects by way of the working equation for a Type 3 Energy-Budget System:

$$\begin{aligned} & a_S S_i + a_L L_i && \text{(heat load)} \\ & \quad + \varepsilon\sigma T_c^{\,4} + H + E + B && \text{(vertical dissipative flux)} \\ & \quad + (H_h + E_h) && \text{(advective dissipation)} \\ & \quad + M_b + M_{Hp} + (M_H + M_E)_a = 0. && \text{(stored and utilized)} \end{aligned} \qquad \textbf{(12-1)}$$

The control practices may, to a certain extent, be grouped according to the principal portion of the energy budget at which they are aimed. Such a grouping forms the primary organization of the discussions to follow.

Controlling the Heat Load: Heat Trapping

By taking proper account of solar geometry, one may direct the flow of the major source of environmental heat—sunlight—so as to produce significant increases in the early season and nocturnal temperatures of the plant environment. Proper direction of the flow may also have beneficial results in utilization of light energy per se, along with the increases in temperature. It should be noted at the outset that, while increases in the heat load in the early part of the growing season may be beneficial when temperature optima for most crop plants are higher, midsummer increases may have extremely adverse effects. Management practices, then, must take account of these changing requirements.

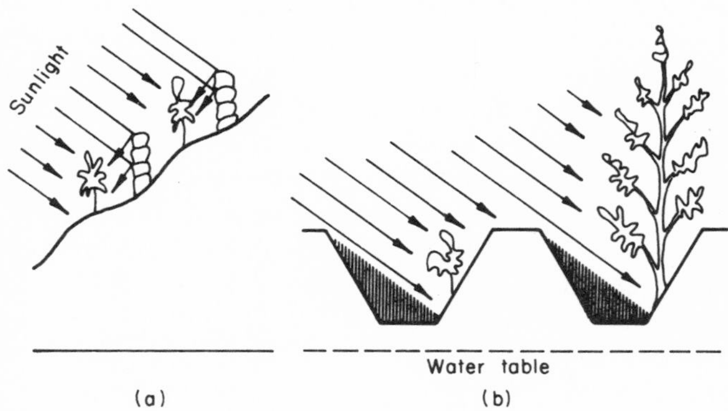

Fig. 12-1. (a) Heat trapping by stone walls between crop rows. (b) Heat trapping by furrow planting. This becomes an intermittent control as plants grow.

As shown in Fig. 12-1a, alternate rows of low stone walls and plants grown on a steep and sunny slope will reflect light to the otherwise shady side of the plants without in turn shading the plants uphill from each wall. In addition, the large thermal capacity of the granite-like materials of such a wall would produce a ready local source of heat by night. On the other hand, without proper precautions, the walls may act as dams impounding cold air on their upper faces—cold air which would otherwise flow in a shallow layer downhill away from the plants. This form of heat trapping is practiced in the vineyards along the Rhine River where the high value of the crop and the absence of any excessively high temperatures in midsummer make this a beneficial continuous control practice.

Another form of heat trapping may be employed in spring for the benefit of young plants by planting them on the sunny walls of furrows, as shown in Fig. 12-1b. The soil itself acts here as reflector and source of nocturnal heat, and again one must be alert to the adverse effects of cold air collecting in the depressions. An interesting added effect is that, in drier climates evaporation losses from the soil may be reduced by having the source of soil moisture near the surface beneath the plants but farther from the surface beneath the intervening ridges. Furthermore, a "continuous" control method becomes an "intermittent" one simply by the growth of the plants to a height much greater than the depth of the furrows in the hotter portion of the growing season.

Both the stone walls and the furrows result primarily in increases in the S_i terms of the energy budget: S_{di} and S_{ui}, but mostly S_{hi}. The results desired are the accompanying increases in plant temperatures, and thus growth, represented by M_b and M_{Hp}. It may be expected that all the other terms in the energy budget will be increased in magnitude as well, this being either beneficial or adverse depending upon other considerations. The effect of these controls at night is to make the magnitude of L_i larger, thus requiring a higher T_c—and thus warmer plant temperatures on a cold night—even though the amounts of energy on an absolute basis may be small compared with those involved during the day.

Controlling the canopy structure of plants by spacing and thinning may be considered a form of heat trapping in that, with attention to proper orientation during midday, light which would otherwise be reflected away by the upper canopy reaches the lower canopy directly and by reflection from the soil. While for an individual leaf this effect is seen in the energy budget as an increase in S_i, for the canopy as a whole the internal reflections are seen as an increase in a_S. The effects on L_i will probably not be substantial at night, but the beneficial increases in M_b by increased total photosynthesis and growth may be realized provided the accompanying increases in T_c and E during the day do not result in a reduction in net PSN and undue moisture stress. An additional benefit from an opened canopy is the enhanced flux of CO_2 downward into the sun-drenched canopy.

While opening a canopy may be a continuous form of heat trapping (in this case as light), recent trials have shown the benefits of what may be employed as an intermittent practice. Covering the soil between plant rows with a white plastic sheet, shaped so as to direct rainfall toward the plants, reflects light to the lower canopy and also reduces evaporative loss from the soil.[(1)] Though yields in corn appear to have been increased both by increased net PSN and by reduced water stress, the increases due to the former were much greater than those due to the effects on the water budget of the plants under the conditions of the experimental work. As noted, if advisable the sheet could be removed easily, making it an intermittent control.

Controlling the Heat Load: Shading

As suggested above, there may be occasions, during times when uncontrolled heat loads are high and plant temperature optima are not, when negative controls on the heat load may be beneficial. Thus, shading by various means is a common control practice for plant environments. Overhead or sunnyside structures employing opaque materials of one kind or another—usually wood or fiber in some form—will reduce S_i by day, with beneficial decreases in T_c and E, and will likewise increase L_i by night, with accompanying increase in T_c. In brief, the overhead materials act as thermostats in the same physical sense as do clouds, but may be applied or removed at will according to the requirements of the moment.

Protection from Cold: Supplementary Heat

While the nocturnal effects of the several heat trapping schemes serve to protect plants from the cold by the retention of heat in the canopy, adding supplementary heat to the environment is also a common practice. An extreme example of an intermittent practice, with attendant high costs and the necessity for careful consideration of the net values of the effort, supplementary heating is used only on high-valued crops and is accomplished in several ways.

Perhaps the least expensive form of this emergency measure is the flooding of the crop, a common practice in low-lying, frost-prone cranberry bogs, where the nature of the site itself greatly increases the probability of freezing temperatures on an otherwise only marginally cold night. This greater frequency of the emergency, of course, makes a less expensive countermeasure all the more attractive. Water is diverted into the bogs and kept there by retaining dikes on all sides. The large thermal capacity and the enormous latent heat of fusion of water make it very unlikely the crop will suffer low-temperature damage, even in the event of an "advective frost" caused by the macroscale movement of a cold air mass into the region. The heat budget of the canopy becomes that of a shallow pond, and the large positive value of advected heat, H_h, which then becomes canopy heat, M_{Hw} (see Chapter 7), completely dominates the budget. When the threat of low temperatures is eased and the water is released from the bogs, the cranberries usually face a slow drying out at essentially wet-bulb temperatures. Though it does not have a particularly adverse effect on cranberry, the low-temperature drying period following removal of the water may be particularly harmful to other kinds of crop plants. Clearly, the necessity for large dikes would make flooding impractical for any but low-growing species.

"Radiation frost" results from rapid, longwave radiant heat loss under conditions of clear skies, dry air, and calm winds at night. Clouds and evaporative cooling the previous day make values of T_c that much lower at sunset and the likelihood of frost damage that much greater. Such situations pose a threat most frequently after the last advective freeze of spring, when plants have begun to grow new and tender foliage and flowering parts.[2] More short-lived, and representing a shortage of heat only in a shallow layer of air near the ground, these radiation frosts are more amenable to countermeasures than the massive assault of an advective freeze.

In another control practice which makes use of the large heat of fusion of water, cold water is applied as a fine spray to plant parts which have already fallen to temperatures slightly below freezing.[3] So long as just the right amount of water is applied and actually freezes on the plant parts, the heat of fusion and the addition of thermal capacity to the plant parts will retard cooling and may prevent the parts from reaching damaging low temperatures before sunrise. If too little spray reaches the plant parts, any movement of dry environmental air will lower the plant materials to the wet-bulb temperature and actually "undercool" them. If too much spray is added, the addition of mass to the plants in the form of ice will so increase the surface/volume ratio that radiative and convective cooling of, for example, fruits exceeds the addition of the heat of fusion. The extra mass may even cause fracture damage to the overweighted structural parts of the plant. Since the cost of hardware for such a practice far exceeds that for ordinary sprinkler irrigation—all parts of the cropped area must be sprayed at once—and since the application rate is so critical, sprinkling for protection against cold is employed only sparingly. Clearly, it is effective only against a radiation frost and not against advective cold spells.

Burning various fuels within the canopy is a frost protection practice in widespread use by orchardists.[4] Given a forecast of critically low temperatures due to radiative heat loss, the crop managers ignite an appropriate fraction of the fuel units which had been placed in the orchard beforehand. The fraction may be increased or decreased through the night as conditions dictate. The fuels burned are most commonly oil, wood slabs, briquettes, or "prestologs." The burning adds heat to the environment both radiatively through an increase in the L_i term and convectively through the M_{Ha} term. Among the increases in the other terms of the budget are the ones in T_c and later in M_{Hp}. While it is also true that the smoke produced by the burning tends to retard radiant cooling to the cold night sky, the contribution of this factor relative to the effects on L_i and M_{Ha} has not been satisfactorily assessed. It is obvious that this practice may be underemployed, but it would seem difficult to burn so much fuel as to cause plant damage. Excessive smoke remaining in the environment the following morning, however, may retard warming and invite considerable criticism from the public.

A sophisticated method for protection of high-value plants against radiative frost makes use of the shallowness of the cold inversion layer of air. Warmer air from above the layer is blown downward into the canopy by propellers on revolving mounts atop towers, or by helicopters moving slowly about the protected area just above the canopy. While the cost of either may be high, the method has the advantages of cleanliness, immediate effect, and the absence of any undercooling from evaporation. In the energy budget, it is the positive increase in the H term translated into an increase in M_{Ha} and M_{Hp} which is the desired effect. Again, it may be possible to underemploy the method but probably not possible to overemploy it, since there are limitations on the ability of the

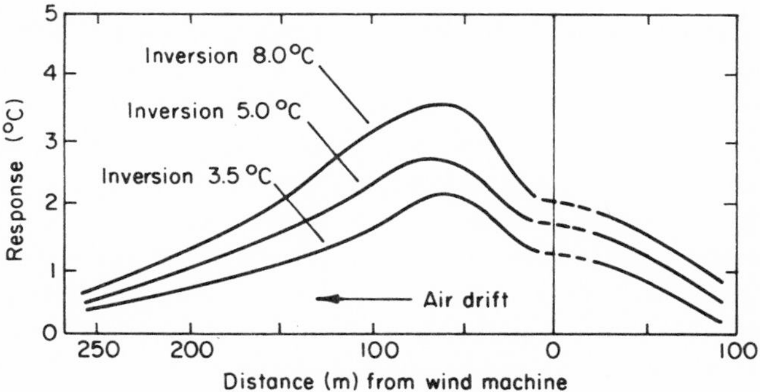

Fig. 12-2. Response (increase) in air temperature as the results of propeller-driven advection, as a function of inversion intensity and distance from the wind machine. (After Wang,[5] from data by Schultz.)

wind machines to move the warm air effectively, as well as limitations on the amount of warm air available.[5] Figure 12-2 shows something of these limitations and the effects of very light wind drift on the distribution of heat by the wind machines. The single wind machine employed in obtaining these results was 90 hp, mounted 32 ft above the ground, and pointed down at a 7° angle below the horizontal. The intensity of the temperature inversions involved is expressed as the temperature difference between 5 and 40 ft above the soil surface in the orchard.

Under some circumstances, it is economically feasible to introduce supplementary heat by way of the soil heat term, *B*, through use of electrical heating cables buried in the root zone of the crop. As with the technique of wind machines, heating cables have the advantage of being rather easily automated to turn on and off at a signal from a proper thermostatic sensor. Producing a positive increase in the *B* term by day or night would probably result in such beneficial effects on plant parts above ground as increases in L_{ui} and M_{Ha}, with resultant increases in M_{Hp} and T_c, despite a likely increase in heat loss due to evaporative cooling at the soil surface.

Protection from Cold: Reduction in Sensible and Latent Heat Flux

Another commonly practiced technique for protection of plants from cold is a combination of several ideas already discussed and a bridge to ideas to be discussed presently. Covering small and young plants—such as peppers and tomatoes—with clear plastic enclosures, or "hot caps," is an effective measure against cold damage in late spring. While the plastic may act thermostatically on the heat load in the same way as a thin cloud, probably its most effective mode of action is through decreased transpirational cooling, *E*, and convectional cooling, *H*, by both day and night during the period of the plant's developmental cycle when temperature optima are highest (see Fig. 11-7). Reductions in the dissipative modes *H* and *E*, of course, necessitate a relative increase in the radiant mode, $\varepsilon\sigma T_c^4$, which involves the increases in plant temperature desired.

It may seem a contradiction at first to say, on the one hand, all the vertical dissipative fluxes of Eq. (12-1) are temperature sensitive and increase with increasing plant and soil surface temperatures, and on the other hand to say we can increase $\varepsilon\sigma T_c^4$ by decreasing H and E. The former statement refers to the situation when the convective exchange area of the plant and the free water surfaces of the mesophyll are in direct contact with the free atmospheric environment. The use of "hot caps" reduces the convective exchange area to that of the cap and removes the free water in the mesophyll from direct contact with the air in the larger environment. While this method has not been applied commercially for protection of larger plants, so far as is known, its advantages of cleanliness, relatively low cost, and application time intermediate between permanence and emergency may make it commercially attractive in the future.

Controlling Soil Moisture: Mulching and Plowing

A great many management practices employed in crop environments operate primarily on terms in the energy budget other than the heat load. Mulching and plowing, for example, both reduce the magnitude of the E term. The reduction is accomplished through the introduction of a quick-drying vapor barrier as well as by a reduction in transpiration from weed species. The vapor barrier is also a thermal insulator, and there is an accompanying reduction in the B term of the heat budget. These reductions are compensated for by midday increases in the magnitudes of both the sensible heat flux, H, and the back radiation, $\varepsilon\sigma T_c^4$. If the mulching material has a lower albedo than the soil it covers, there may be large increases in convective and radiant losses upward by day. If the albedo of the mulch is higher, on the other hand, there may be only relative increases in H and $\varepsilon\sigma T_c^4$—as compared with E and B—and very small absolute increases. As may be seen in the discussion below concerning results obtained with sheet mulches, the effects of mulch albedo on the energy budget are undoubtedly affected by the size and density of the canopy involved.

What mulches are used? One form of mulch is a blanket of particles spread over the soil surface. The results of shallow plowing produce such a mulch, but sawdust, wood chips, straw, and various other materials are used where plowing is not appropriate. A second form of mulch used is the sheet—usually of plastic—which may be translucent, dark, or light. A third type so far developed takes the form of a monomolecular film, a petroleum gel, or some other impervious fluid which assumes the shape of the surface it is mulching.

The interplay between a particle mulch, or shallow plowing, and the remainder of the physical environment offers a perfect example of a control practice which is definitely a mixed blessing. While it is true the practice reduces the water losses from a crop system, it also effectively removes the soil as a heat reservoir on cold nights, thus increasing the frost hazard to plant parts above ground. During advective cold, on the other hand, it may protect plant parts underground from prolonged cold weather. As it keeps the root zone from cooling too rapidly, it may also retard root warming when young plants have higher temperature optima in spring. Particle mulches may introduce toxic materials to the soil while, if they are organic mulches, they may reduce the amount of available nitrogen in the soil by the requirements of their own decomposition. Particle mulches do a marvelous job of reducing wind and water erosion of surface soil and of retaining the water from light rainfall in the immediate environment. By the same token, they are likely to keep soils cold and wet in seasons of unexpectedly high precipitation.

These counterbalancing advantages and disadvantages suggest the need for "intermittent" mulching, which in many respects is the great advantage of a plastic sheet mulch: it may be put down and taken up to some extent on short notice. Of course, particle mulches may be blown onto a planted area, as is often the case on newly seeded highway cuts, but they are rather difficult to remove later when the crop plant has begun to grow. The same is true, to an even greater extent, of a gravel mulch sometimes used in ornamental horticulture and landscaping.

Sheet mulches may be perforated to prevent concentration of precipitation in a small fraction of the soil area. In regions of sparse growing season precipitation, however, this concentration may be just what is needed. The experiments cited above, on the use of white sheet mulch in corn (see "heat trapping"), indicate that, while white sheets produce greater yields by comparison with black sheets, because of reflected light reaching the lower canopy, the white and black produced very small differences from one another in soil temperature. In the heat of the growing season, however, these mulches both reversed the tendency to act as thermal insulators and produced distinctly warmer soils by comparison with uncovered plots, because of reductions in evaporative cooling at the surface. Translucent plastic mulches, what is more, produce even warmer soils, not only be reducing evaporative cooling, but also by acting as a greenhouse over the soil. The effect on destructive soil microflora of unnaturally great soil warming produced by these sheet mulches has not been thoroughly investigated.

Research on the use of surface films has only recently begun, but it appears to offer many of the advantages and fewer of the disadvantages of other mulches. In particular, the reflective and absorptive properties may be predetermined in the commercial mixing process, and the longevity—and thus the intermittency—may be controlled simply by knowing the deterioration rate of the film materials and setting the type and volume of film material applied to the surface.

Beginning with uniformly wet soil, the film mulch produces higher temperatures and greater moisture retention at seed depth in the first few days of drying.[(6)] The added warmth continues throughout a daily cycle but is accentuated at midday. Water moves from the warmed surface layer, as both liquid and vapor, into the seed layer a centimeter or two beneath the surface. The result is a soil column with a greater mean temperature and a water content not only retained but also concentrated at seed depth. The effects of the mulch through longer periods and through several wetting cycles have not been examined. Thus, such effects as those of increased warmth and wetness on microfloral pathogens remain to be studied.

Controlling Soil Moisture: Irrigation

Perhaps the oldest and most widely used artificial control of plant environments is irrigation. While it assuredly reduces moisture stress on the growing crop, it also tends to keep the E term as large as the drying potential of the air will allow, thus keeping the plant environment relatively cool, especially in the root zone. In certain portions of the growing season this may be a distinct advantage to the crop. It may also be an advantage given by the control practice to soil disease and pest organisms. As with mulching, irrigation as a control on the E term may be a mixed blessing. It certainly has the advantage of intermittency, and with careful attention to timing and rate of application of water, irrigation has long ago proved itself a beneficial control practice for plant environments.

Controlling the Soil Heat Budget

Although mulching and irrigation have incidental effects on the B term of the energy budget, there are practices employed on occasion which are intended to have their primary effect on the B term. Orchardists regularly roll and pack the soil between trees after they have cultivated to remove weeds. The packing reduces the adverse effects of the insulator left by loosening the surface soil, while at the same time conserving moisture and changing the albedo relatively little. The desired effect, then, is achieved on marginally cold nights when the soil heat reservoir makes the difference between frost damage and no frost damage.

Although the striking effects on the B term of changing the albedo of the soil surface have been demonstrated experimentally (see Case 6, Chapter 4), the commercial potential seems not to have been exploited. Something of these effects was suggested above in discussion of the mixing of film mulches, but neither films nor powders for albedo control seem to have been employed as such in crop management.

Although the burning of crop residues after harvest in perennial seed grass—the reverse of mulching—is a control practice aimed primarily at elimination of pathogens and pests, it does affect the energy budget. The fire, in removing less healthy portions of the grass plants, opens the canopy and leaves deposits of unburned carbon and ash. These materials decrease the albedo, and very likely contribute to soil warming, which in turn increases growth of the crop following the fire.

Controlling Convection and Advection: Shelter Belts

A control technique which operates directly on the advective and convective portions of the energy budget is the interplanting of a tall plant species among the individuals of a shorter, more valuable species. The plantings in the form of rows of hedge or trees or both around the edges of fields are known as "hedgerows" or "shelterbelts." When the two species are scattered essentially evenly over the planting area, the taller is known as the "companion crop" or "nurse crop." In either case, the effect of the taller species is to brake wind action at the level of the shorter one, thus reducing turbulent transport of heat and moisture upward from the cropping. The same braking on the wind by plants upstream will reduce advection of various properties to or from the site of concern.

The exact effects of secondary plantings on the terms of the energy bugdet depend upon the geometrical relationships between the two species. For example, a large shelterbelt greatly reduces the wind speed near ground level when it lies perpendicular to the windflow—about five tree heights upstream, and ten to fifteen tree-heights downstream. Figure 12-3 suggests the pattern of these effects in a vertical plane perpendicular to a shelterbelt. The effect of such a shelterbelt on the heat load of nearby fields is probably very minimal except in the areas currently in shade. With a reduction in H, E, and advective terms over sunny areas, and no reduction in the heat load, the responses of the system are increases in both back radiation and soil heat flow: increases in plant temperatures above and below the soil surface.

The situation is rather different for nurse croppings. Here the heat load on the shorter, primary plants would be reduced by shading of the taller ones in the immediate environment. The effects on the other energy-budget terms would be difficult to

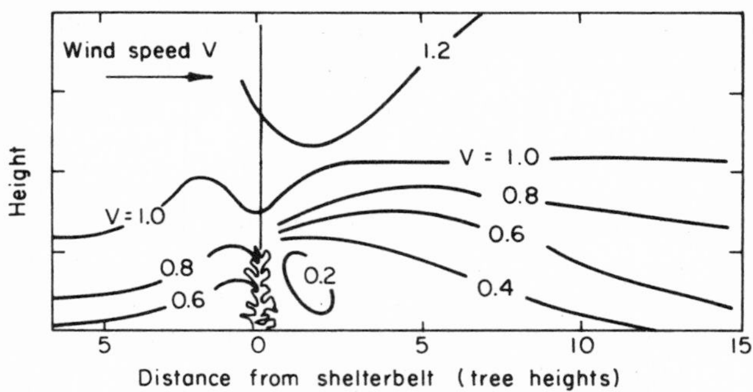

Fig. 12-3. Generalized pattern of wind speed reduction in the neighborhood of a shelterbelt. (After Fig. 266 of Geiger,[9] from data by Nägeli; and Woodruff and Zingg.[10])

determine without additional knowledge, again of the differences in plant height and spacing. Measurements might show, for example, that all terms of the heat budget are reduced along with the heat load in the case of a very dense nurse crop and a short primary crop. A less dense nurse crop might not reduce the heat load nearly so much, with results on the other terms more nearly like those of a shelterbelt.

Shelterbelts in the Great Plains area of the central United States have an interesting history.[7] The idea that they should be planted was promoted during the 1930's on the basis of the contention they would remove the threat of drought from the area by pumping soil moisture from the soil into the air through the transpiration of the trees, there to be returned by rainfall to the fields between the shelterbelts. The large losses of soil moisture from the areas near the trees was recognized, but seen as an ultimate benefit to agriculture nearby. The extra transpired moisture, however, was blown away by the winds and never realized as additional precipitation on the fields. As meteorologists know, what caused previous droughts was not the lack of atmospheric moisture, but the lack of proper atmospheric processes to form clouds and precipitation.[8] The shelterbelts only increased water loss from the deeper soil layers but had no effects on the larger-scale precipitation-forming processes.

The failure turned into a success, however, when the beneficial effects already mentioned were found in nearby fields. In addition, the shelterbelts greatly reduced windborne soil erosion, collected windborne snow near fields and kept it from being blown (advected) into stream beds, and provided—as the Europeans had found many years before—a wonderful habitat for wildlife near the fields. Properly placed and thinned, the shelterbelts would not act as cold air dams, and on balance, the use of these plantings has been found to be a blessing in the Great Plains. The positive aesthetic effects on the appearance of the plains and the reduction in heating requirements of sheltered farm homes (Type 4 systems) are additional fringe benefits.

The question of what effects man's activities have on the local water balance is a never-ending one. It is argued, for example, that under certain circumstances extra water sources may indeed produce a net increase in precipitation nearby. There is little doubt that, as mentioned, shelterbelts and small manmade lakes cannot create precipitation by adding water vapor to the local air mass. There is reason to believe, however, that they may occasionally so humidify the air beneath rainclouds that some of the precipitation, once formed, will reach the surface when it would otherwise have evaporated as it fell through dry air. Extra water caught in a rain gauge nevertheless may not represent an increase of water cycling beneficially in the local soil-plant-atmosphere system, nor any improvement in the time of arrival of rain as an aid to crops. Thus, the argument goes on.

Concluding Remarks

Table 12-1 presents a cursory review of the various techniques for artificial control of plant environments. Clearly, the battery of practices available to the plant husbandman may be used in various combinations so as to optimize the physical and economic results. For example, plowing followed by soil packing is a combination regularly used in orchards. Heat trapping in spring followed by shading in summer is popular with nurserymen.

Doubtless there are other combinations, and even other practices, which have not come to the attention of plant scientists. The

Table 12-1
SUMMARY AND CRITIQUE OF GENERAL TECHNIQUES FOR ARTIFICIAL CONTROL OF PLANT ENVIRONMENTS

Control practice	Primary effects on...	Major beneficial results of factors	Major adverse results of factors
Heat trapping	$a_S S_i + a_L L_i$	Higher plant temperatures and PSN in spring	Higher plant temperatures in hot weather
Shading	$a_S S_i + a_L L_i$	Lower plant temperatures in hot weather	May be expensive if employed intermittently
Flooding	M_{Hw}	Frost protection	Retarded warming after cold threat
Spraying cold water	M_{Hp}	Frost protection	Very expensive; undercooling with improper application rate
Burning fuels	M_{Ha}	Frost protection	Expensive; remaining smoke retards warming
Wind machines	H	Frost protection	Very expensive
Soil heating cables	B	Frost protection	Very expensive
"Hot capping"	E and H	Higher plant temperatures in spring	Very expensive
Mulching and plowing	E	Reduces water loss; frost protection for root zone; reduced erosion	Increased frost hazard above ground; possible soil toxicity
Irrigation	E	Reduces moisture stress	May benefit soil-borne pests; retards soil warming
Soil packing	B	Reduces frost hazard	
Shelterbelts and nurse crops	H and E	Reduces losses of heat, moisture, and erosion in spring	Soil moisture losses from deeper layers

day may come when control practices now considered too intensive may be found economically beneficial even in range, forest, and "wildland" management. One of the keys to any successful control program, however, would seem to be a thorough knowledge and appreciation of the energy-budget concept.

REFERENCES

1. "Reclaiming sunlight," *Agr. Res.* **14** (8), U.S. Dept. of Agriculture, Agr. Res. Service, February 1966.
2. See for example, N. J. Rosenberg and R. Myers, *Monthly Weather Rev.* **90**, 471 (1962).
3. A brief but useful account of the theory of this practice may be found in J. A. Businger, "Protection from the Cold: Frost protection with Irrigation," *Meteorol. Monograph* **28**, 74, Amer. Meteorol. Soc., Boston, Massachusetts, 1965.
4. R. Geiger, "The Climate near the Ground," p. 509, Harvard Univ. Press, Cambridge, Massachusetts, 1965, describes some results of this practice and presents a bibliography of European writing on the subject.
5. Figure 11-2 is from J. Y. Wang, "Agricultural Meteorology," p. 550, 1967, who gives another useful bibliography on these control methods. Further discussion, including a theoretical treatment, may be found in T. V. Crawford, "Protection from the Cold: Wind Machines and Heaters," *Meteorol. Monograph* **28**, 81, Amer. Meteorol. Soc., Boston, Massachusetts, 1965.
6. A. Kowsar, M.S. Thesis, Department of Soils, Oregon State University, Corvallis, Oregon, 1968.
7. See, for example, A. H. Carhart, "Shelterbelts: the failure that didn't happen," *Harper's* p. 75 (October, 1960).
8. An interesting and humorous commentary on this subject may be found in J. E. McDonald, "The evaporation-precipitation fallacy," *Weather* **17**, 168 (May 1962).
9. R. Geiger, "The Climate near the Ground," p. 501 (4th ed.), Harvard Univ. Press, Cambridge, Massachusetts, 1965.
10. N. P. Woodruff and A. Zingg, *Trans. Amer. Geophys. Union* **36** (2), 203–208 (1955).

PROBLEMS

12-1. Suppose you planted crop rows two plant-heights apart rather than the one plant-height apart you had used in previous years. (One plant height is the height of an average mature plant of that species.) What would be two effects of your new planting scheme on the physical environment, and what would you expect as a physiological consequence of each effect?

12-2. Examine the effects of heat trapping in the following way, using the experimental results of Table 8-8. Assume for the windy site that trapping increased the the total incident shortwave energy and the total incident longwave energy each by 10%. Then calculate the amount of increase in leaf temperature of cherry for each wind speed if the transpiration rate is not changed.

12-3. Examine the effects of "hot capping" in the following way, using the experimental results of Table 8-8. Assume the environmental conditions are those of the windless site (which was actually in a greenhouse), and that the small air volume within the cap produces a 50% reduction in E and in H. What will be the resulting temperature of an oak leaf in the cap when equilibrium is reached?

12-4. Examine the effects of orchard heating in the following way. The trees stand 5 meters tall and occupy 2% of the volume below that height. Assume the thermal capacity of live wood and leaves is 0.5 and that of air is 25×10^{-5} cal cm^{-3} °C^{-1}. You generate heat by burning residual fuel oil in burners which each produce 36×10^{6} cal/hour, but 60% of that heat is lost from the plant-air system by various processes. How many burners per unit of ground area do you need to produce at least a 1.0°C/hour increase in the temperature of the air and the tree materials in the orchard?

Chapter 13

ANIMALS AND THE ATMOSPHERE

As HAS BEEN the case with our examination of atmospheric effects on plants, so in this chapter we will deal with only enough of the effects on animals to build a foundation for understanding and to suggest the variety and complexity of these effects. While some of the similarities to analyses of plants will be recognized, it is the differences which characterize animals that may be of more interest to the reader.

Direct and Indirect Atmospheric Effects on Animals

As with plants, the atmosphere may affect the lives of animals both directly and indirectly. Direct effects, in turn, may be viewed as either short-term or long-term. Short-term effects of the environment pertain to the ways in which the comfort and survival of individuals are determined by the interactions of energy transfer and metabolism between animal and environment. Long-term effects are those in which the integrated short-term effects exert their control on development of individuals and lead to successful or unsuccessful survival and reproduction of a population of individuals.

Indirect effects may be viewed, to use an electrical analogy, as series effects or parallel effects. Indirect series effects refer to matters of the animal's position in a food chain. The environment determines in large measure whether the organisms lower in the food chain will be available at the right times to provide the quantity and quality of food necessary for the survival and reproduction of the organism of concern. In addition, the presence of organisms higher in the food chain, acting as predators, is greatly influenced by the environment. Indirect parallel effects on the organism of concern pertain to the presence and effectiveness of diseases, parasites, and symbionts. We shall be dealing only with the direct effects—both short-term and long-term.

Short-Term Effects: The Preferendum

Unlike plants, animals are mobile and have the potential for seeking out, from among a variety available, a microenvironment which in some respect best meets their needs of the moment. If this choice is freely expressed in the presence of a complete spectrum of environments, the microenvironment selected is termed the "preferendum." In a sense, the preferendum is analogous to the optimum of a plant in that both tend to represent conditions which are somehow "best" for the organism. To the extent that a "comfortable" environment may be less than optimally "healthy," however, the preferendum departs from the optimum. It should not be difficult for one acquainted with life in alternately air conditioned and uncontrolled surroundings to imagine that what is most comfortable may not be most healthy. The possibility of such a departure would seem to be a meaningful objective of biometeorological research with animals.

Field conditions may afford a sufficiently restricted variety that a choice made by a free-ranging animal may not actually represent a true preferendum. However, it is likely that a choice made in a complex natural landscape of stream, forest, and meadow, for example, would be nearer to a preferendum than one made in shortgrass prairie. Thus, the analog of providing artificially controlled environments for plants (Chapter 12) would appear to be the providing of a wider choice of microenvironment for animals. Such a wider choice is that suggested in the discussion of shelterbelts and hedgerows with regard to wildlife in the last chapter. The controlled laboratory environments for animals represent the other extreme, while the slightly greater variety in a barnyard is something in between.

In discussing the physical meaning of an organism's preferendum, it is useful to consider the working equation of a Type 4 energy-budget system:

$$a_s S_d A_{sd} + a_s S_e A_{se} + a_L L_i A_{se} \quad \text{(heat load)}$$

$$+ \varepsilon\sigma A_{se} T_s^4 + L_v K(e_m - e_a)/(R_m + R_a) + h_c(A - A_c)(\Delta T) + kA_c(dT/dz) \quad \text{(dissipative fluxes)}$$

$$+ M_b + M_{Hp} = 0. \quad \text{(stored and utilized)}$$

(13-1)

This is essentially a rewriting of Eq. (8-16), except that the E and the M terms have been expanded for convenience of discussion. The E term is similar to that for the transpiration rate in Eq. (10-6). Vapor losses by breathing and integumental flow form a more complex system than transpiration, but we may still represent it by a vapor pressure difference, $(e_m - e_a)$, divided by a series resistance, $(R_m + R_a)$. Here m refers to conditions within the organism and a to conditions at the organism-atmosphere interface. L_v is again the latent heat of vaporization for water, and the vapor exchange area for the organism is included within the coefficient of proportionality, K. The stored and utilized energy is divided into the two components of the chemical energy in the animal, M_b, and the sensible heat content of the mass of the animal, M_{Hp}.

An animal's seeking its preferendum may be viewed as a behavioral means of controlling its energy budget. Thus, movement is one of the several modes of behavioral energy regulation, along with posture, ingestion, and architecture. In seeking a sunny or a shady, a windy or a calm, a moist or a dry environment, the animal selects a combination of heat load and dissipative fluxes, usually a combination with an equilibrium temperature within a fairly narrow range. In assuming a certain posture, the animal also modifies the combination by changing the areas A_{sd}, A_{se}, and A_c. By ingesting cold water, for example, an animal loses some body heat in warming the water to body temperature. Technically, this ingestion must be considered as a small change in the definition of the system involved, unless an equal amount of water is shortly lost or dispelled from the body. At any rate, ingestion of food and water not at body temperatures may be viewed as a means of behavioral thermoregulation.[1] Construction of various nests, beds, etc. beyond the mere selection of a site must be viewed as an architectural means of behavioral thermoregulation apart from movement per se.

In addition to the voluntary behavioral modes of thermoregulation, there are also involuntary modes. Panting, sweating, opening spiracles, etc. are responses to overheating which take the form of an increase in K and a decrease in R_m in the latent heat term of Eq. (13-1). The result is an increase in evaporative cooling. Involuntary responses to cold stress also take various forms in higher forms of animals. Shivering converts energy to heat through the performance of work. Further, the contraction of blood capillaries near the skin reduces the volume of blood flowing near the organism-environment interface. This reduces the skin temperature, and thus decreases (dT/dz) in the conductive dissipaton term of Eq. (13-1). The reverse of this response—the dilation of capillaries and the increase in (dT/dz)—is an additional involuntary response to overheating.

As noted, the combination of responses an animal makes to its physical environment usually results in the attainment of an equilibrium body temperature which lies within a narrow range. In Eq. (13-1), the condition is represented by a steady-state value of zero for M_{Hp}. Figure 13-1 gives a feeling for the nature of this equilibrium in the form of a frequency distribution of individual fish which, given free choice on a broad spectrum of water temperature, moved to the temperatures indicated in the relative proportions shown.* Three-quarters of the experimental fish expressed temperature preferenda falling within a range of only 3°C. Since for an aquatic the energy budget consists largely of only the H and the M terms, we may assume the water temperatures and the body temperatures for these fish were essentially the same.

* Unless otherwise noted, the figures and examples discussed in this chapter are adapted from citations in Andrewartha and Birch.[2]

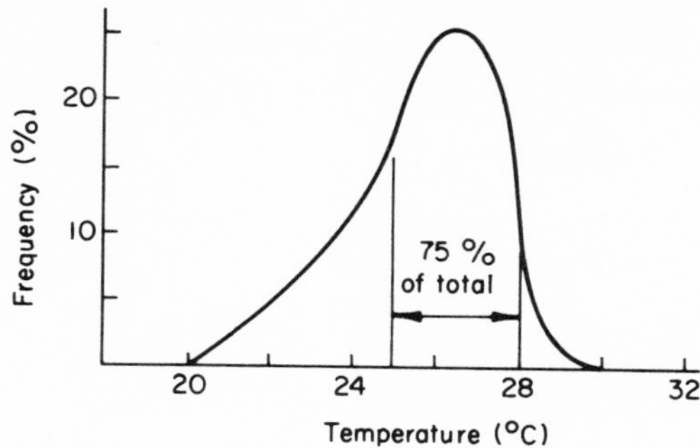

Fig. 13-1. Relative frequency with which temperatures were sought by fish given a wide spectrum of choice in an experimental channel.

Thus, the narrow range of equilibria mentioned is illustrated directly by this example.

The shape of the distribution curve—with a tailing off toward lower temperature—is intriguingly similar to frequency distributions of the body temperatures of reptiles measured in the field. The similarity may be coincidental, however, since, as Heath[(3)] points out, the observed temperatures of the reptiles do not necessarily represent behaviorally regulated results. He reached this conclusion after obtaining a similar distribution by measuring temperatures of water-filled beer cans at randomly selected times on a sunny day. The temperatures he obtained are clearly just the equilibria of the inanimate Type 4 systems, and the point is well made that analyses of field data should be approached with caution. The easy and tempting transfer from a true preferendum such as that of Fig. 13-1 to a similar interpretation for the field data Heath describes must be avoided.

Wellington's work with spruce budworm larvae[(4)] provides an example of a preferendum observed in terrestrial rather than aquatic animals, which also contains a lesson in data interpretation. Groups of larvae were placed in test chambers, some chambers with gradients of relative humidity only and some with gradients of both temperature and relative humidity. The larvae moved so as to seek temperature-moisture preferenda as shown in

Fig. 13-2a. The preference being expressed was for neither temperature nor relative humidity, as is clear when measurements of "evaporation rate" from water-filled micropipettes in the chambers

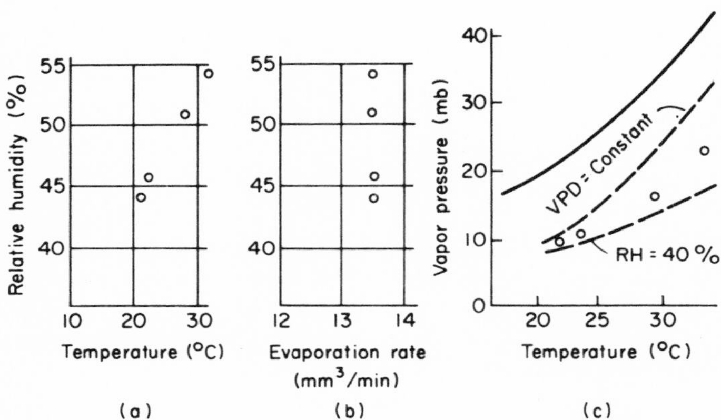

Fig. 13-2. (a) Combinations of temperature and relative humidity sought by larvae of Spruce budworm in an experimental chamber. (b) Constancy of evaporation rate from micropipettes at the locations sought by Spruce budworms. (c) Temperature-moisture preferenda on a *TRe* diagram, together with lines of constant vapor pressure deficit and constant relative humidity. Data from Wellington(4) as in Table 13-1.

are used as a coordinate in Fig. 13-2b. But why would an insect larva prefer conditions characterized by water loss rate from a water surface?

As discussed previously, evaporation rate from a water surface and the vapor pressure deficit in the air next to that surface are linearly related so long as the water and the air are at the same temperature. One might suspect, therefore, that Wellington's larvae would be expressing a preference for a particular VPD since their small bodies might be nearly at air temperature. But Wellington's calculations show the behavior was not clearly related to VPD, but to the "evaporation rate."

Table 13-1 describes the physical conditions at the larvae's preferred locations, including values taken directly from the

Table 13-1
NUMERICAL DESCRIPTION OF EXPERIMENTAL TEMPERATURE-MOISTURE ENVIRONMENTS SOUGHT AS PREFERENDA BY LARVAE OF SPRUCE BUDWORM[a]

Wellington's data:	Experiment number			
	1	2	3	4
Dry-bulb temperature, T (°C)	20.6	22	28	32
Relative humidity, R (%)	44	46	51	54
Vapor pressure deficit (mm Hg)	10.2	10.7	13.8	16.1
Evaporation rate (mm^3/min)	13.5	13.5	13.5	13.5
Further calculations:				
Wet-bulb temperature, T_w (°C)	13.5	15	20.5	24.5
Saturation vapor pressure at T, e_s (mb)	24	26.5	38	48
Saturation vapor pressure at T_w, e_w (mb)	15.5	17	24.5	30.5
Ambient vapor pressure, e (mb)	10.7	12.4	19.2	25.6
VPD ($e_s - e$) (Same as Wellington's VPD)	13.3	14.1	18.8	22.4
VPD ($e_w - e$) (mb)	4.8	4.6	5.3	4.9

[a] Basic data from Wellington.(4)

original paper. The results of further calculations show that the VPD obtained using the wet-bulb temperature as the water temperature is nearly constant. Thus, it may be the animals' bodies were near the wet-bulb temperature and they were expressing a preference for a certain rate of water loss. Though the data are too imprecise to make an exact physiological inference, it is clear the preference has more to do with water regulation than with temperature.

Whether Wellington's findings are valid under field conditions is not known, but the lesson on data interpretation is clear: Had he been guilty of single-factor thinking and observed only temperature, say, he would have concluded the larvae had no preferendum in the environment provided. Incidentally, he observed in an extension of this work that the evaporation rate preferred decreased steadily as larval age increased beyond the second instar of the animals depicted in Fig. 13-2 and Table 13.1.

As just noted, factors other than species determine the value of the preferendum for a particular atmospheric variable. Besides age, preconditioning of the animal is one of these factors. The fish whose temperature preferenda are expressed in Fig. 13-1 selected lower preferenda if the temperature to which they had become accustomed before testing was low. The nature of this dependence is shown in Fig. 13-3, where it is clear the temperature preferendum

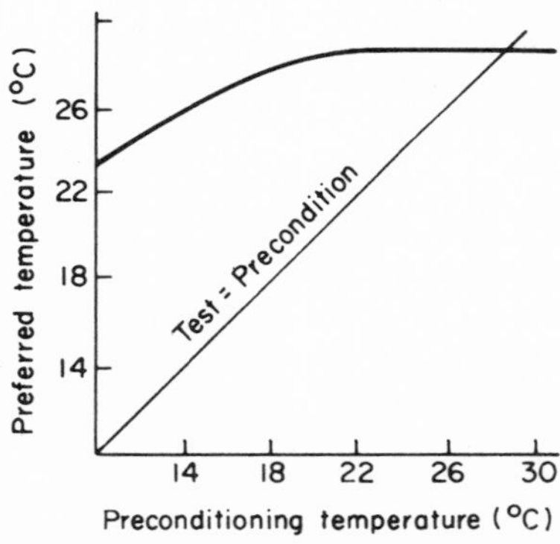

Fig. 13-3. Temperature preferenda of fish as a function of their preconditioning temperature.

is dependent on preconditioning temperature only, below a certain point. Thus, the tailing off of the frequency distribution in Fig. 13-1 is doubtless due mainly to effects of preconditioning, while the tailing off of the distribution of temperatures in Heath's beer cans (and by his inference also that of reptiles' temperatures measured in the field) is due to an observation schedule which consisted mainly of times when the heat load was near the maximum for the day and less frequently of times with a submaximal heat load.

Alteration of the moisture preferendum by preconditioning to various levels of physical well-being is seen in Fig. 13-4. In this

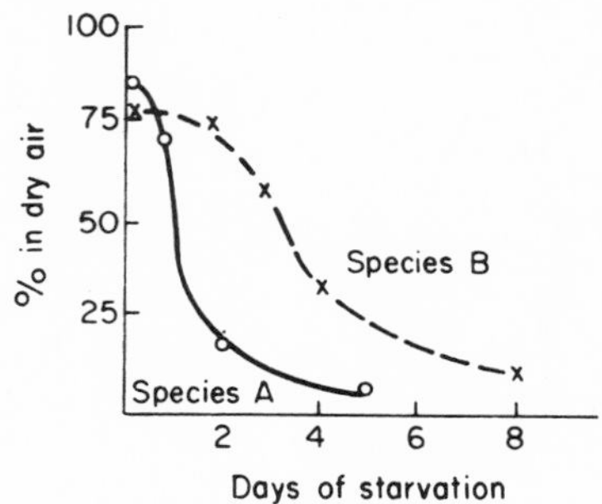

Fig. 13-4. Moisture preferenda of two species of beetle as a function of length of preconditioning time in a dehydrating environment without food.

experiment, beetles were restricted to an extremely dry environment without food or water for various numbers of days. They were then released into an apparatus which gave them free choice of two environments: one slightly more moist than that of preconditioning (called "dry") and one very moist. The proportion choosing the same dry environment as the control animals in the experiment fell as the number of days of treatment increased. While it seems reasonable to suppose the preference was for the conditions which most quickly provided "rehydration," it is interesting to note the species differences in the reaction to stress.

To this point we have been concerned with the concept of a preferendum, its meaning in terms of the energy budget of an animal, voluntary and involuntary responses to stress, and alteration of responses by preconditioning. As the physical environment exerts considerable control over whether or not an animal seeks a different environment, and over which environment he seeks, so it also exerts control over his ability to reach it.

Short-Term Effects: Rate of Dispersal

As one could expect from knowledge of the relationships of temperature and rates of energy-converting biochemical reactions, the speed of movement of animals—and by inference the rate of dispersal—increases to a maximum at some optimum temperature and then decreases at higher temperatures, as shown in Fig. 13-5a.

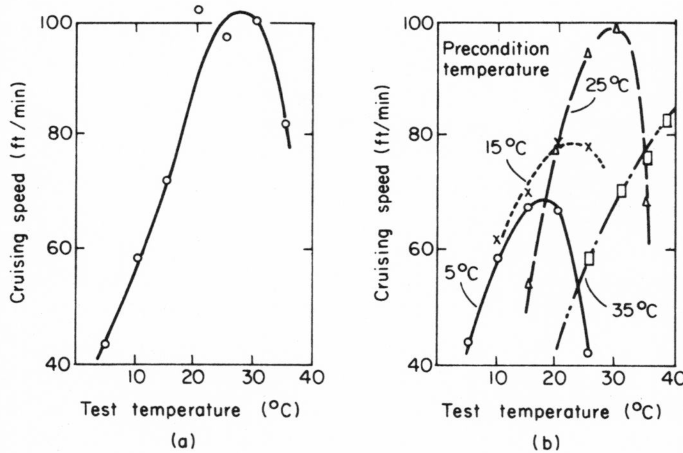

Fig. 13-5. (a) Maximum 2-min cruising speed of goldfish in an experimental channel, as a function of testing temperature. (b) Maximum cruising speed as a function of testing temperature and preconditioning temperature.

Here measurements were made of the fastest speed a goldfish could maintain for 2 min in an experimental channel held at the temperature to which the fish had become accustomed before the test. The interesting effects of temperature preconditioning may be seen in Fig. 13-5b. These results were obtained from fish tested at temperatures different from their accustomed temperatures and show clearly that, within the range of experimental variables employed at least, animals whose metabolism is adjusted to lower temperatures perform best at relatively low temperatures. The question of whether this effect is due to physiological regulation or to genetic control is discussed for cattle by Johnson.[1] The question is certainly an interesting one in these days when individuals of the various races of man are increasingly often taking up residence in climates very different from those of their ancestors.

It may be instructive in closing consideration of direct short-term environmental effects on animals to reassert the complexity of the problem of analysis. One may wish go take account, for example, of effects on preferendum or rate of dispersal of the spatial density of the population being observed. It seems reasonable that in the field, and certainly in a small laboratory, the fish seeking a preferendum will move to and be tallied at a temperature different from their true preferendum if the density of other fish is already too great at the location of that true preferendum. Thus, the variability shown in Fig. 13-1 may be in part due to the crowding of fish in a small apparatus as well as to the effects of preconditioning. It is likely this effect of crowding would be felt only in the extreme case of a very restricted area, but its relevance to data analysis should be borne in mind all the same.

Andrewartha and Birch[2] argue most convincingly that effects of population density are felt in the action of every environmental variable which controls the numbers of the population. They argue equally well that the rate of dispersal, as measured by observed dispersion rather than by such a maximum as shown in Fig. 13-5, is relatively independent of density. Thus we remind ourselves that the biological environment of an organism modifies in subtle ways the organism's reactions to the physical environment, with perhaps the added subtlety of reactions being different in natural conditions than in experimental laboratory conditions.

Long-Term Effects: Rate of Development

In Chapter 11 a generalized relationship between environmental temperature and a plant's rate of development was discussed. Many experimental data for animals suggest the same form of relationship, as may be seen for data on fruitfly eggs in Fig. 13-6a. Here the time required to complete the egg stage is plotted

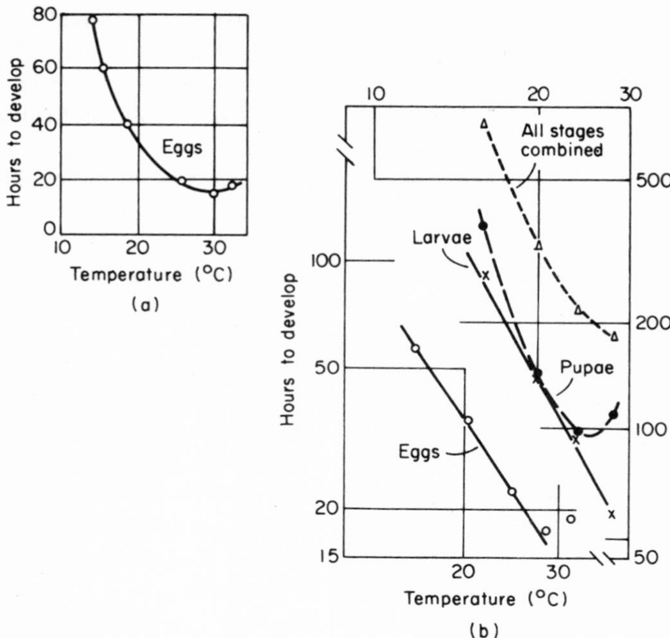

Fig. 13-6. (a) Number of hours for fruitflies to pass through egg stage, as a function of constant environmental temperature, on linear coordinates. (b) On double logarithmic coordinates, number of hours for fruitflies to pass through egg, larval, and pupal stages.

on linear coordinates against the temperature of the environment in which they were reared experimentally. This curve is very clearly similar to that in Fig. 11-9 and the same remarks concerning the morphological meaning of the curve made in Chapter 11 apply here.

Figure 13-6b shows the same time-temperature relationship for eggs plotted on double logarithmic coordinates, together with the relationships for the same experimental insects in the larval and pupal stages of development. Several things may be noted from these data as plotted. First, for the egg stage and the larval stage, logarithmic presentation produces nearly a straight line in the portion of the temperature scale between about 16 and 28°C. This linearity disappears on the same part of the temperature scale for the pupal stage, and therefore also for the time for development from oviposition of the eggs through the pupal stage. For such relationships which do appear linear, the empirical equation for the data is

$$D = a/T^b, \tag{13-2}$$

from which the inverse relationship of the development time, D, and the development temperature, T, is obvious.*

Second, the temperature optimum for the egg stage which appears near 30°C is found nearer to 24°C for the pupal stage. This is in qualitative agreement with the lowering of temperature optima with advancing age in young plants, as shown in Fig. 11-7. As also mentioned in the discussion of the heat units concept, which has often been applied to entomological problems, the neat relationships suggested by such equations as (13-2) are only approximately correct. The tempting simpliciy of the HU concept becomes even more misleading when the time for more than one developmental stage is under consideration, as is clear in Fig. 13-6b. The facts remain, however, that rate of development in plants and animals is temperature sensitive and that the overall relationships between rate and temperature are amenable to approximate analysis and application.

The discussion of temperature and rate of development to this point has been concerned only with experimental results obtained with constant temperatures. As with plants, again, temperature fluctuations may produce markedly different results and must be considered more representative of natural conditions. It may be that animals—more than likely small ones—are thermoperiodic in

* As Andrewartha and Birch[2] (pp. 144 *et seq.*) point out, a better empirical equation for most such data is

$$\ln(K - D)/D = a - bT,$$

which is semilogarithmic and in which a, b, and K are experimental constants. They also point out any effort to give a mechanistic explanation for the form of either equation is probably unwarranted.

the same sense as plants. On the one hand, it is certain that short exposures to extreme temperatures cause permanent impairment of normal development in some animals. On the other hand, such exposures produce no aparent effect in other animals, despite the fact that the extreme temperatures would be lethal if exposures were prolonged.

An even more striking kind of result comes from seasonal, rather than diurnal, temperature fluctuations. There is considerable evidence that various forms of arrested development, for example successful survival adaptations enabling individuals to endure seasonally harsh weather of some sort, are hormonally controlled. The form is known as "hibernation" in larger animals and "diapause" in insects. In the case of diapause at least, the phenomenon not only permits survival under harsh conditions, it is even obligate to the extent that individual insects which have not undergone diapause develop quite differently from those of the same species which have. An example of this is shown in Fig. 13-7.

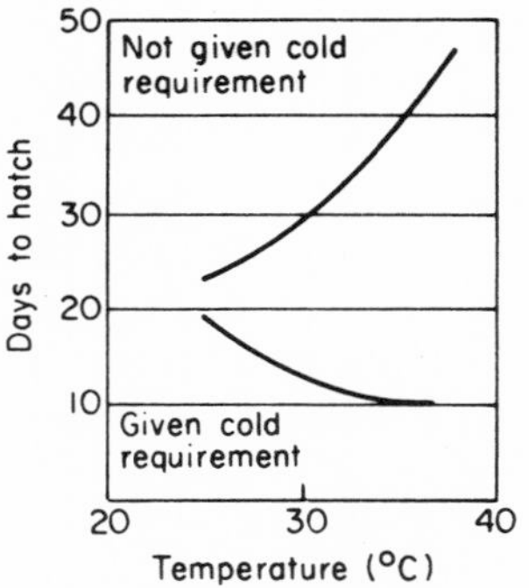

Fig. 13-7. Number of days for grasshopper eggs to hatch as a function of hatching temperature and of exposure to cold requirement of diapause.

The harsh conditions which appear necessary in this instance are certain exposures to low temperature. The insect eggs given a prolonged exposure to low temperature responded to subsequent exposure to constant temperatures as indicated by the curve marked "given cold requirement." This is the sort of response considered normal in light of discussions above. Eggs of the same species not given the exposure to cold, on the other hand, showed exactly the opposite response to constant temperatures: Higher temperatures reduced the rate of development even in the temperature range considered favorable for the species under normal circumstances. It will not be possible to consider the many known variations in timing and intensity of these seasonal temperature effects on animals. Suffice it to say they must be taken into account in any analysis of short-term, as well as long-term, responses of animals to environment.

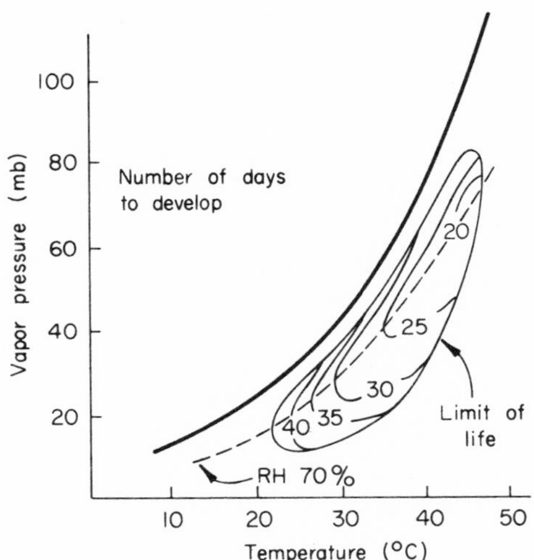

Fig. 13-8. Number of days for locust nymphs to reach adulthood as a function of temperature and moisture. Data from Hamilton.[5]

To illustrate the necessity for taking other environmental variables into account in analysis of the rate of development, Fig. 13-8, again from the entomological literature,[6] shows rate of development for locust nymphs as a function of temperature and moisture. On a *TRe* diagram, one may see that the number of days required for the locusts to become adults is strongly dependent upon both variables. The question of what constitutes an optimum again arises here in that the combinations of temperature and moisture which produce a requirement of only 20 days for development are very near the outer limits for life according to Hamilton's experiments. As one might conclude on the basis of intuition, rapid development is not necessarily healthy.

The research represented in Fig. 13-8 took account of the possibility that fluctuating environmental conditions might produce different reactions than those to the constant environments shown in the figure. Within the range of the observations, the data showed only minor differences between constant and fluctuating environments when the latter were characterized as weighted mean temperatures and relative humidities. For example, an environment programmed to alternate a 7-hour cycle at 38°C and 60% with a 17-hour cycle at 20°C and 90% was characterized as an environment of 25°C and 81% and produced results agreeing very closely with those from constant conditions shown in Fig. 13-8. The interesting thing is that life for these animals would have been impossible in a constant environment of 20°C and 90%, but the 7 hours in the warmer, drier environment permitted life to be sustained. This observation is in support of the assertions made above about the necessity of considering fluctuating conditions when extrapolation of laboratory data to the field is considered.

Again in Fig. 13-8, we see that the rate of development shows its greatest values (least number of days) near 70% relative humidity irrespective of temperature or vapor pressure. In view of the contents of Table 13-1 and the discussion about them, it is likely that the physiological basis is the same for both the preferendum of spruce budworm larvae and the rate of development of locust nymphs, and that the basis has a great deal to do with the rate of water loss from the insects' bodies. It is interesting to note that the "chosen" relative humidities of the mesic budworm seem to be distinctly lower than those of "optimal" development for the desert-inhabiting locust.

It will not be possible to do more than mention the fact that light is known to have definite, but poorly understood, effects on the long-term processes in animals. Such a statement would apply equally well, it seems, to plants.

Long-Term Effects: Reproduction

Figures 13-9 and 13-10 and a comparison of them with Fig. 13-8 will disclose at once the simplicity and the complexity of the subject under discussion. The three figures are all based upon the

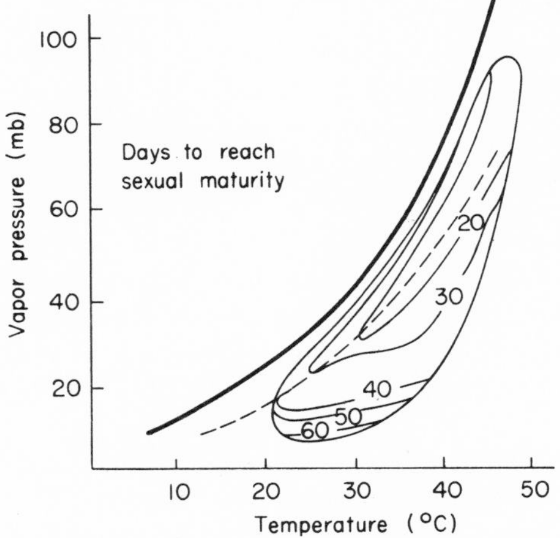

Fig. 13-9. Number of days for new adult locusts to reach mating age, as a function of temperature and moisture. Data from Hamilton.[5]

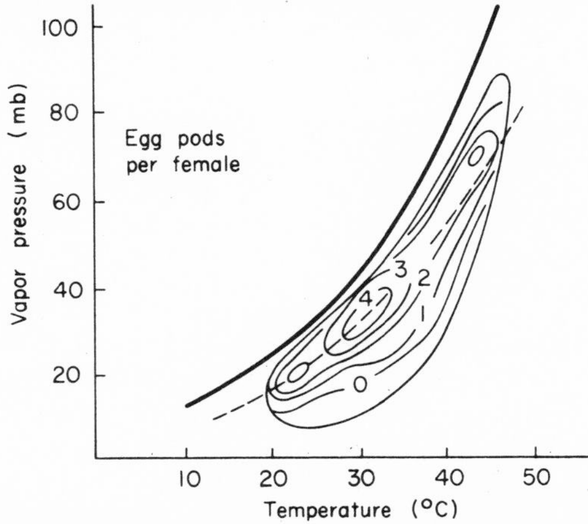

Fig. 13-10. Number of egg pods laid per female locust, as a function of temperature and moisture. Data from Hamilton.[5]

same experimental work by Hamilton[5] and all have a similar appearance, suggesting a quasidependence upon relative humidity per se as the controlling variable in the physical environment. Together these figures also exhibit the complex sequence of physiological and phenological events which is continuously dependent on the conditions of the physical environment.

Figure 13-8 shows the dependence for nymphs reaching adulthood; Fig. 13-9 shows the dependence for adults reaching sexual maturity; and Fig. 13-10 shows environmental effects on a measure of the first crucial step in reproduction: the number of egg pods per female. To this point we can trace the "constancy" of dependence at each step on some factor closely related to the water budget of the insect. Hamilton's work does not enable us to trace moisture effects on the subsequent stages of incubation and hatching, but his temperature data in Table 13-2 make it abundantly

Table 13-2
RELATIONSHIPS OF INCUBATION TIME AND HATCHING SUCCESS TO TEMPERATURE FOR A SPECIES OF LOCUST[a]

	Temperature (°C)						
	24	25	27	39	32	38	43
Incubation (days to hatch)	22.3	19.2	15.0	13.8	12.9	12.6	12.2
Nymphs hatched per pod	26	44	42	30	28	25	14

[a] Data from Hamilton.[5]

clear these stages continue under the critical environmental control of the lengthy and complex process of reproduction leading to the net result we call "fecundity": the number of living offspring per adult. The important implications of the relatively minor changes in dependence on the environment experienced during the life cycle of a small animal are directly comparable to those discussed above in relation to plant development.

Most of the examples in this chapter have come from the literature on fish and insects, because these examples are most readily available and because they make the important points so well. What of extrapolation of the ideas developed here to other, distinctly different animal life forms? Suffice it to say that, although most of the ideas probably apply at least qualitatively to these other animals, too-strict extrapolation can be misleading. As one simple, direct example of this, Hamilton found that a maximum value of the response measure "nymphs per pod" (last line of Table 13-2) was not to be found within the same temperature range in work with another very similar species of locust.

Concluding Remarks

The frequent references to plant responses to the environment in this chapter on animal responses is a purposeful attempt to suggest strong connections between plant ecology and animal ecology often overlooked. In addition, it is certainly no accident that most of the experimental data come from plant and animal forms which are small, relatively simple, and otherwise convenient to work with.

It may be instructive here to review some of the connections to be seen between this chapter and Chapter 11. It is clear that both plants and animals are linked to the environment through direct, indirect, short-term, and long-term effects. While plants may not be said to exhibit movement in search of a preferendum, it would be meaningful to note the similarity between postural thermoregulation in animals and the wilting and heliotropic responses of some plants. Voluntary or involuntary, they all control the area factors in Eq. 13-1, and thus play important roles in the energy budgets of individual organisms. While plants do not appear to have a response analogous to an animal's ingestive controls on its energy budget, it may not be amiss to suggest there is an analogy between an animal's architectural behavior and a plant's morphological adaptations.

Despite quantitative differences in the relationships between temperature and rate of development, plants and animals appear strikingly similar in their qualitative responses to the environment. This similarity, of course, makes any successful techniques in the general application of the heat unit concept likely to be transferable from one life form to another.

Finally, it should be clear from previous discussions that the environment exerts a continuous but ever-changing control over the series of developmental, or phenological, steps leading to reproduction in both plants and animals. True understanding will come only with knowledge of the physiological processes which are linked to the environment and which produce the quantitative changes with age in the responses to the environment.

It is becoming increasingly clear that the division of physiological ecology into botanical and zoological phases is inhibiting progress toward true understanding. In reading the next chapter, the reader may well conclude that our further division into a human phase of physiological ecology is similarly difficult to justify.

REFERENCES

1. These matters of thermoregulation and the effects of heat stress on metabolism and productivity in cattle and other domestic animals are discussed very instructively in Harold D. Johnson, "Response of animals to heat," *Meteorol. Monograph* **28**, 109, Amer. Meteorol. Soc., Boston, Massachusetts, 1965.
2. H. G. Andrewartha and L. C. Birch, "The Distribution and Abundance of Animals," Univ. of Chicago Press, Chicago, Illinois, 1954.
3. J. E. Heath, *Science* p. 784, (Nov. 6, 1964).
4. W. G. Wellington, *Sci. Agr.* **29**, 201 (1949).
5. A. G. Hamilton, *Trans. Roy. Entomol. Soc. London*, **101**, 1–58 (1950).

PROBLEMS

13-1. You plan to conduct an experiment similar to Wellington's (Fig. 13-2 and Table 13-1) but with larvae of a different species. You know from the literature that your larvae consistently seek an environment in which the wet-bulb vapor pressure deficit $(e_w - e)$ is 8 mb. If you put a number of larvae in a chamber controlled everywhere to a relative humidity of 30%, but with a range of temperatures, to what temperature will you expect the larvae to move?

13-2. Given the relationship below between mean daily temperature and the number of days for an organism to complete development through the phenological subperiod in question, estimate from the climatological data given the day on which development will be complete if it begins on Day 1.

Mean daily temp (°F)	40	50	60	70	80	90	100
Days to develop	14	8	5	3	2	2	3

Day No.	1	2	3	4	5	6	7	8	9	10
T_{max}	60	65	70	80	75	60	70	80	90	100
T_{min}	40	45	50	50	55	40	40	50	50	60

Chapter 14

HUMANS AND THE ATMOSPHERE

IN MANY WAYS this chapter is just an extension of the previous one. Broadly, all the remarks concerning short-term and long-term, direct and indirect, parallel and series effects of the atmosphere on animals hold true for human biometeorology. Man, along with most other animals, employs posture, ingestion, movement, and architecture as voluntary modes of thermoregulation in seeking a preferendum. Likewise, along with other animals, man is equipped with such involuntary modes of thermoregulation as panting, sweating, shivering, variable pulse rate, and variable rate of blood circulation near the surface.

In this chapter, the ways in which humans differ most from other animals will be emphasized, but at the outset certain matters common to all animals must be considered. Certain principles of biophysical adaptation to environmental stress must be examined, and man put into perspective, before his most striking adaptational tools—clothing and architecture—are discussed. In the course of this chapter virtually no mention will be made of indirect or long-term atmospheric effects. Nearly the entire chapter will deal with man's short-term responses to environmental stress.

Biophysical Adaptation: Effects of Animal Size

As a model of a homeothermic animal, consider a spherical container filled with water kept at a constant temperature. As noted in Eq. (13-1), the various net rates of energy exchange between such a model and its surroundings are proportional to the surface area of the model, which area is in turn proportional to the square of the sphere's radius, r^2. That is to say, the rate of energy expenditure which must be maintained to hold the model's temperature constant in a cooler environment is proportional to the surface area and to r^2. Since the model's volume is proportional to the cube of the radius, r^3, it follows that the rate of energy expenditure for the homeothermic model is proportional to the two-thirds power of the volume, and for constant density of the model, the mass: $(\text{mass})^{2/3}$. This relationship is qualitatively true for a model of any shape and for any homeothermic organism and its effective radius, surface area, and mass. The relationship is pictured in Fig. 14-1, where it may be seen at once that an animal

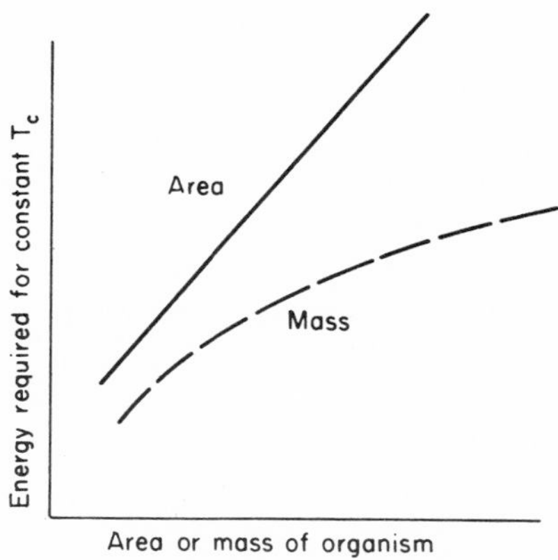

Fig. 14-1. The energy required for a constant core temperature in a homeotherm is proportional to its surface area, and proportional approximately to the 2/3 power of its total mass.

may be of a relatively large mass, or size, and still have a relatively small metabolic requirement. For example, while a mouse requires 200 Kcal/kgm of body weight, an elephant requires only 8 kcal/kgm.[(1)]

Also related to the surface and the mass of an organism is its "time constant." A body without temperature regulation, when introduced into an environment at a different temperature, will approach the temperature of the environment in such a way that, at any given time, the rate of temperature change in the body will be proportional to the body-environment temperature difference at that time. The equation for such a response curve is the "logarithmic decay"

$$d(\Delta T)/dt = (-1/\theta)(\Delta T) \quad \textbf{(14-1)}$$

where ΔT is the body-environment temperature difference at time t, and θ is the "time constant" of the body. The response curves for typical large and small bodies placed in a cooler environment are shown in Fig. 14-2. The same relationships, inverted, would apply

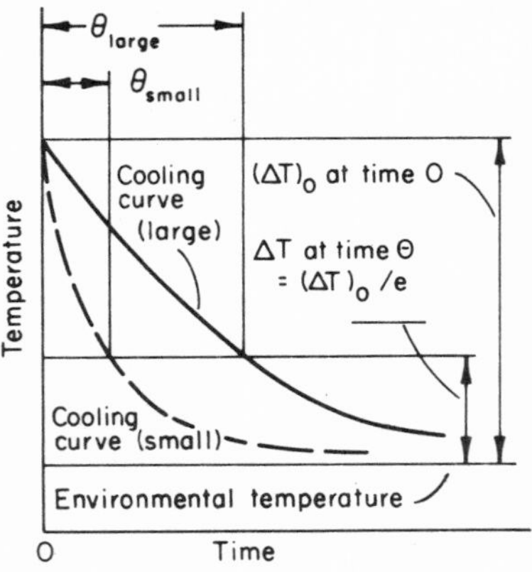

Fig. 14-2. Schematic cooling curves for large and small masses, illustrating the meaning of the time constant for the masses.

for cool bodies in a warm environment; and one sees that θ is small for a small mass and large for a large mass. In fact, θ is the amount of time required for the temperature difference, ΔT, to be reduced to $1/e$ (i.e., about 1/3) of its original value. Since for a given mass, the value of θ will also be in inverse proportion to the surface area of convective exchange, the time constant is a reflection in one number of the combined effects of both mass and surface. While a mercury thermometer may typically have a time constant around 1 min, and one of Heath's water-filled beer cans a time constant around 1 hour (see Chapter 13), a water-filled 55-gal drum may have a time constant around 1 day. The implications for short-term response of an organism to its environment are obvious.

Biophysical Adaptation: Physiological Responses

In light of the discussion of the effects of animal size on conductive-convective heat exchange with the environment, it should be clear that smaller homeothermic animals would be likely to evolve a dependence upon dynamic, quick-responding mechanisms of insulation from the physical environment, while larger homeotherms would appear with more static responses. To borrow the phrase of Herrington,[1] meteorologically defined distinctions of weather and season may be biologically inappropriate, since "in a stimulus sense what is weather for one animal may be season for another."

In small homeotherms, one finds the seasonal appearance and disappearance of peripheral fat and hair coats supplementing the short-term temperature dependence on hair coat erection and metabolic rate as prime biophysical responses. As shown in Fig. 14-3, this temperature responsiveness of metabolic rate amounts to about 1.67 kcal/m^2/hour/deg deviation from a 28°C thermal neutrality in the rat but to virtually a zero metabolic response in

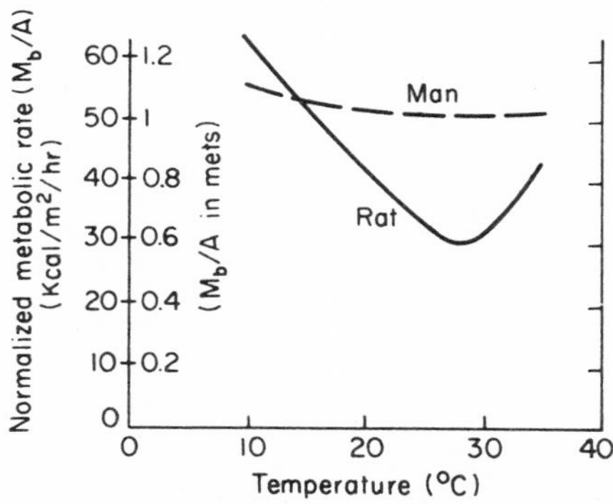

Fig. 14-3. Normalized metabolic rate of heat production for a rat and for a man as functions of environmental temperature. After Herrington.[1]

humans. It is interesting to note in passing the several figures in Chapter 13 in which temperatures near 28°C appear to be optimal in some sense. This temperature also constitutes a form of thermal neutrality in man, as will be seen presently.

Hairless man depends upon other responses to thermal stress. Hyperthermal stresses result in sweating, in increased pulse rate with attendant circulatory rate, and in vasomotor dilation of blood capillaries near the skin. Hypothermal stresses result in the reversal of these responses. The characteristics of the vasomotor response are shown in Fig. 14-4, where the effective thermal con-

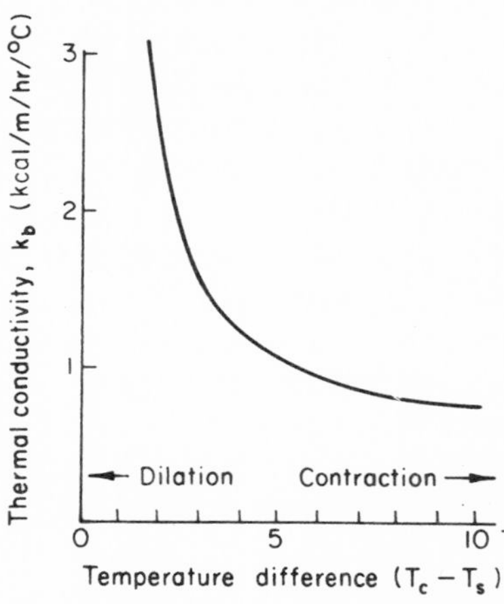

Fig. 14-4. Effective thermal conductivity of a man's body as a function of the core-skin temperature difference, as calculated from Eq. (14-2) using data from Herrington.[1]

ductivity of the body and its circulatory system are shown as a function of the temperature difference between the body core and the skin. The values of this conductivity were obtained from the equation:

$$k_b = L_b(M_b + M_{Hp})/A_{se}(T_c - T_s) \qquad \textbf{(14-2)}$$

where T_c is the core temperature (usually rectal), T_s is a representative skin temperature, L_b is taken as 0.1 meter—a representative value for the path length between core and skin—and the other symbols are as in Eq. (13-1).* The manner in which k_b increases with the dilation of capillaries under heat stress is clearly shown, the increase having the physical result of a greater heat transfer between core and skin for a given temperature difference. The increased pulse rate also plays a role in this increase in k_b. The contraction of capillaries, as noted in the previous chapter, results in a larger temperature difference between core and skin but a smaller difference across the skin-air interface where convective dissipation of body heat takes place.

Not a great deal has been said up to this point about the algebraic signs to be associated with the components of the M term in Type 3 and 4 energy budgets. Careful examination should reveal that M_b logically has a positive sign when there is a net rate of heat production from biochemical energy conversion within the system. In animals and humans, therefore, M_b should always be positive. M_{Hp} should logically have a positive sign associated with a withdrawal of heat from the mass of the system, and therefore with a fall in the mean temperature of the system's mass. M_{Hp} may be thought of here, then, as a positive withdrawal rate.

Beyond man's "emergency" responses to thermal stress—secretory, vascular, and circulatory—are his most flexible and powerful response modes: clothing and architecture. It is with these that most of the remainder of this chapter will deal.

The Energy Budget: Operative Temperature

In order to appreciate the approach biometeorologists have taken to problems of the energy exchange between humans and the environment, begin by rewriting Eq. (13-1) as follows:

$$
\begin{aligned}
& a_s S_d A_{sd} + a_s S_e A_{se} && \text{shortwave} \\
& \quad + \varepsilon\sigma(T_r^4 - T_s^4)A_{se} && \text{longwave} \\
& \quad + L_v K(e_m - e_a)/(R_m + R_a) && \text{evaporative} \\
& \quad + h_c A_{se}(T_a - T_s) && \text{convective} \\
& \quad + M_b + M_{Hp} = 0, && \textbf{(14-3)}
\end{aligned}
$$

* Since conceptually this thermal conductivity is represented by an equation like (4-2), the equation Herrington gives is doubtless in error.

where the incoming longwave term is now represented by a fourth power emission at a mean radiant temperature of the environment, T_r; the emissivity of body and environment is taken as a single value, ε; the convective area is assumed equal to the radiational area A_{se}; and the conductive term connecting body and substrate is assumed zero.

It is often convenient, once the temperature T_r has been defined, to derive a radiant transfer coefficient, h_r, so that one may write

$$\varepsilon\sigma(T_r^4 - T_s^4) = h_r(T_r - T_s).$$

The coefficient h_r is only slightly dependent on the values of temperature involved, but its use requires expression of temperatures on the absolute scale.* Net longwave radiant transfer and net convective transfer may then be written in similar terms of the form $h(\Delta T)$. With the assumptions that $S_d = S_e = 0$, that evaporation is negligible, and that thermal equilibrium makes $M_{Hp} = 0$, solving Eq. (14-3) for the surface temperature, T_s, yields

$$T_s = \frac{M_b}{A_{se}(h_c + h_r)} + \frac{h_r T_r + h_c T_a}{h_r + h_c}. \tag{14-4}$$

The first term on the right is nearly a constant, and the second term on the right is a weighted average of the air temperature and the effective radiant temperature of the environment.

With (M_b/A) for a man about 50 kcal/meter^2/hour, or about 0.083 cal/cm^2/min, and with h_c and h_r both on the order of 10^{-2} cal/cm^2/min/°K, the first term on the right of Eq. (14-4) has a value of about 4.5. Since the second term on the right will be

* The argument behind the approximation is that

$$(T_r^4 - T_s^4) = (T_r^3 + T_r^2 T_s + T_r T_s^2 + T_s^3)(T_r - T_s).$$

Since on an absolute temperature scale T_r and T_s are not likely to be greatly different, the approximation follows that

$$(T_r^3 + T_r^2 T_s + T_r T_s^2 + T_s^3) = (2T_r^3 + 2T_s^3) = 2(T_r^3 + T_s^3),$$

so that $h_r = 2\varepsilon\sigma(T_r^3 + T_s^3)$.

near 300 using absolute temperatures, the first term is neglected in defining the "operative temperature":

$$T_{op} = (h_r T_r + h_c T_c)/(h_r + h_c). \qquad \textbf{(14-5)}$$

Recalling that h_c is a function of wind speed and h_r weakly related to environmental temperature, one may see that in a variety of experimental situations where S and E are negligible, the effects of various combinations of radiant (wall) temperature, air temperature, and wind speed may all be reduced to the single value T_{op}. The condition that $M_{Hp} = 0$ is simply a statement that the body is in thermal equilibrium with the environment, but the removal of the restrictions on S and evaporation lead to a much more complex, and thus less often used, relationship than the operative temperature so familiar to human biometeorologists.[2]

Lee[3] has proposed an empirical approach to the problem of accounting for the S and the E terms in the calculation, in which measures of physiological response are employed to establish equivalence of two environments. In broad outline, the method consists of four steps:

(1) Expose exercising subjects in a control chamber where $S = 0$ and where T_a and e_a are controlled at various values.

(2) Using values of T_a and some dependent physiological variable, such as sweat rate, develop statistical relationships between the two variables.*

(3) Observe the dependent variable in subjects performing the same exercise in the same chamber with solar radiation present.

(4) Using the relationships in (2), calculate the equivalent operative temperature, T'_{op}, which acting alone with $S = 0$ would have produced the value observed in (3).

The difference between T'_{op} and the T_{op} in the sunlit control chamber of step (3) is the appropriate correction to T_{op} for the observed shortwave radiant heat load on the experimental subjects.

* With relationships $W = A + BT_{op}$, where W is the sweat rate (gm/hour), A and B are constants, and the correlation coefficient of the relationship is between 0.80 and 0.90, Lee calculated $(T'_{op} - T_{op})$ to be about 7.5°C under midday conditions of summer at Yuma, Arizona.

The Energy Budget: Evaluation of Components

Clearly, there is a variety of combinations of wind, radiation, air temperature, and humidity which will produce a given physiological response in a homeotherm such as man. As has been shown, the equivalence of these various combinations can be calculated and expressed in a single number, the operative temperature. In Fig. 14-5 are presented some results from Herrington[1] which show the energy budgets of clothed and unclothed subjects at various operative temperatures. As in other considerations, an energy budget component is assigned a positive value if it represents a flow of energy to the system. In the case of Fig. 14-5, the system is the skin surface of the human subject. Hence, metabolic energy, M_b, which arrives at a constant rate of about 50 kcal/meter^2/hour* from within the system is positive at all times, while stored heat from the body mass itself, M_{Hp}, arrives from within only under cold stress. The rationale for calling M_{Hp} a positive withdrawal may be clearly seen.

Several comments about Fig. 14-5 should make its meaning clear. First, for any temperature, the algebraic sum of the components is zero, in accordance with Eq. (14-3). Herrington has lumped both radiant exchange modes and convection into one term $(R + C)$, leaving E, M_b, and M_{Hp} separate. Second, even at low temperatures, breathing results in a small negative E. Third, the temperature of thermal neutrality near 29°C represents a condition wherein no hyperthermic or hypothermic responses are active, and wherein the addition of light clothing has no effect on either $(R + C)$ or E.† Fourth, as one would expect, $R + C$, being indirectly dependent upon the value of $(T_c - T_s)$, becomes zero near the core temperature, $T_c = 36$°C. Fifth, sweating and the E component are the "ace in the hole" which constitutes the major

* This value for M_b is taken as a standard unit of metabolic heat production in human biometeorology: 1 Met = 50 kcal/meter^2/hour. (See Fig. 14-3.)

† The value of this temperature of thermal neutrality in man and in many other animals is often taken as evidence of their evolution under tropical conditions.

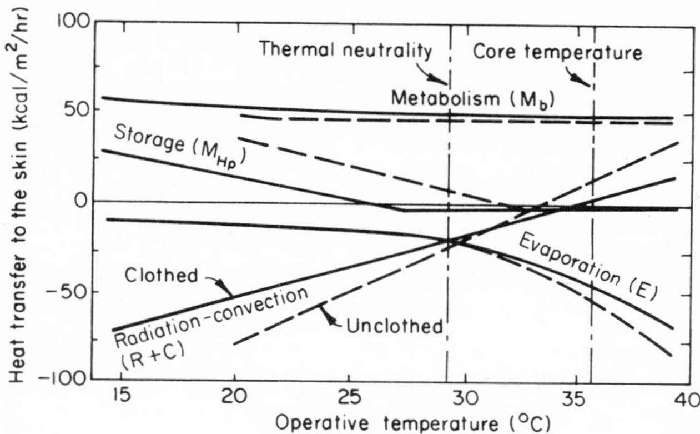

Fig. 14-5. The energy budget of a man, clothed and unclothed, as a function of operative temperature. After experimental results of Herrington.[1]

short-term response to hyperthermal stress which has been so strikingly developed in man. Finally, clothing, as we shall discuss at greater length presently, acts thermostatically to reduce values of both $R + C$ and E at temperature extremes. Clothing also reduces the temperature at which M_{Hp} becomes positive. In this connection it should be said that, of course, the body or any peripheral members of it cannot long withstand conditions which produce positive M_{Hp} without damage to tissue.

The Role of Clothing

Discussion of the role of clothing in the energy budget of a human may be undertaken most directly by means of an equation relating clothing and other factors to the temperature difference between the environment and the body core. Begin with an equation for the energy budget of the outer surface of a layer of clothing, in which we assume $T_a = T_r$, combine the direct and diffuse shortwave radiation terms, and employ the same approximation for net longwave transfer which led to Eq. (14-4):

$$a_s SA_x + A_x(h_c + h_r)(T_a - T_x) + A_x E + A_x(k_c/L_c)(T_s - T_x) = 0 \tag{14-6}$$

where the terms in order are: (1) net shortwave flux, (2) net flux due to longwave radiation and convection combined, (3) latent heat flux, and (4) conduction from within the layer of clothing. The symbols, where different from those used previously, are

S, incoming direct and diffuse shortwave radiation per unit time and per unit area,

A_x, the area of the external clothing surface whose energy budget is expressed by the equation, and taken as a single value for all transfer processes in order to facilitate later discussion.

T_x, the temperature of the outer surface of clothing,

E, the rate of latent heat transfer per unit time and per unit area,

k_c, the thermal conductivity of the clothing, and

L_c, the clothing thickness.

To proceed, we combine (14-6) with an equation describing the conduction across the interface between the skin and the inner surface of the clothing, and with Eq. (14-2) for conduction between the body core and the skin. Flow at the skin-clothing interface is

$$(k_c A_s/L_c)(T_s - T_x) = (k_b A_s/L_b)(T_c - T_s) \tag{14-7}$$

where k_b and L_b are defined in Eq. (14-2). A_s is the total skin area in contact with the inner surface of clothing, and the temperatures of air, skin, and body core are T_a, T_s, and T_c. Obtaining expressions for T_x and T_s in terms of T_c from Eq. (14-7) and (14-2), and substituting them into Eq. (14-6), results in the relationship originally sought between clothing and the core-environment temperature difference:

$$(T_c - T_a) = \frac{(M_b + M_{Hp})R}{A_s} + \frac{(a_s S + E)}{(h_c + h_r)} + \frac{(M_b + M_{Hp})}{A_s(h_c + h_r)} \tag{14-8}$$

where R is the combined "resistance" to conduction by the body and the clothing $(L_c/k_c) + (L_b/k_b)$. We may now examine the various relationships which affect the dependent variable $(T_c - T_a)$, and in particular the role of clothing in making it possible for a homeothermic human to maintain a constant value of T_c.

First, look at the equation in general to see that it makes sense. In a hot environment where T_a is larger than T_c, it is necessary for M_{Hp} to be negative and of larger magnitude than M_b unless the evaporative cooling rate, E, is very large. This simply says that, with only a modest evaporative cooling rate, a large T_a will result in a large negative withdrawal rate and the body will tend to rise in temperature. Notice also that since the M terms are not normalized for area in their definition, they must be normalized here. S and E, of course, are expressed per unit area already. Finally, note the disappearance of A_x in going from Eq. (14-6) to (14-8).

In a cold climate with T_c much larger than T_a, all terms in Eq. (14-8) will be positive. In fact, it is the cleverness and effectiveness with which he kept these terms large and positive that has enabled man to move into such hostile cold environments on earth. How has he done this?

Table 14-1, containing representative values of the resistance R, will permit a direct assessment of the role of clothing. In the table, the body and clothing thicknesses are assumed to be $L_b = 0.1$ meter and $L_c = 0.01$ meter. It is clear from the table that clothing can increase the resistance greatly beyond the increase possible from vasomotor controls alone. That is, with adequate clothing, the vasomotor control of k_b provides only a "fine tuning" on the overall value of R. According to Eq. (14-8), the increased resistance clothing provides in turn results in a larger value of $(T_c - T_a)$. That is, clothing permits maintenance of a given constant T_c in a colder environment.

Sample calculations will indicate something of the physical meaning of Eq. (14-8). The man may be described as having a cylindrical diameter of 20 cm, an albedo of 0.30 ($a = 0.7$), a normalized metabolic rate of 0.083 cal/min with a water loss rate of about 68 cc/hour (-0.033 cal/cm^2/min), and an effective body conductivity of 0.35.

With a convective transfer coefficient of

$$h_c = (2.2 \times 10^{-3})(V)^{1/3}(D)^{-1/2},$$

a wind speed of 10^4 cm/min (about 3.7 mph) makes $h_c = 10^{-2}$.

Table 14-1
INFLUENCE OF BODY AND COAT CONDUCTIVITY ON THERMAL RESISTANCE, *R*

	k_c [a]			
k_b [a]	0.02 (still air)	0.03 (dry wool)	0.10 (wet wool)	10.0 (air in gentle motion; i.e., no clothing)
1.0 (capillaries contracted)	$R = 0.60$ [b]	0.43	0.20	0.10
3.0 (capillaries dilated)	0.53	0.36	0.13	0.03

[a] The units of k in the table are (kcal meter^{-1} hour^{-1} degree^{-1}) = (0.167 cal/cm/min/deg). The values of k_b are inferred from Herrington's version of Fig. 14-4, and the values of k_c for still air and wool are from Buettner.(2) The value of k_c for moving air was chosen to be suitably "large."

[b] The units for the resistance R in the table are (meter^2 hour deg/kcal) = (600 cm^2 min deg/cal). According to Buettner,(2) human biometeorologists take as the standard unit of clothing resistance 1.7 meter^2 hour deg/kcal = 1 Clo.

If $h_r = 0.86 \times 10^{-2}$ and the direct beam solar flux is 1.0 cal/cm^2/min for a solar altitude of $\theta = 30°$, the resulting calculations are as in Table 14-2. The general situation is of a man clad in a wool suit, standing in a gentle wind under midday conditions in a northern locality in the United States.

The calculated difference between the body core temperature and the general temperature of the environment (T_r was assumed equal to T_a) is 32.5°C. With the core temperature of about 36°C assumed for man, the interpretation is that the wool suit enables him to be in equilibrium with an environmental temperature just above freezing. With the sun still shining, but without the suit, the calculation is that equilibrium would be obtained if the environmental temperature were (36 – 18.5) = 17.5°C, since the thermal resistance would be only about one-fifth its former value. Finally,

Table 14-2

CALCULATION OF THE TEMPERATURE DIFFERENCE ($T_c - T_a$) FOR A CLOTHED MAN STANDING IN A MODERATE MIDDAY ENVIRONMENT ($M_{H_p} = 0$) [a]

Basic variables: The environment:	The man:
$S_d = 1.0$ cal cm^{-2} min^{-1}	$D = 20$ cm; i.e., $L_b = 10$ cm
$h_r = 0.86 \times 10^{-2}$ cal cm^{-2} min^{-1} deg^{-1}	$(M_b/A_s) = 0.083$ cal cm^{-2} min^{-1}
$V = 10^4$ cm min^{-1} (about 3.7 mph)	$h_c = 10^{-2}$ cal cm^{-2} min^{-1} deg^{-1}
The clothing:	$a_s = 0.7$
$k_c = 0.005$ cal cm^{-1} min^{-1} deg^{-1}	$k_b = 0.35$ cal cm^{-1} min^{-1} deg^{-1}
$L_c = 1$ cm	$E = -0.033$ cal cm^{-2} min^{-1}

Calculations:

Calculation	Result
$(M_b/A_s)R = (0.083)[(10/0.35) + (1/0.005)]$	$= +19.0°C$
$a_s S/(h_c + h_r) = (a_s S_d \cos\theta/\pi)/(1.86 \times 10^{-2}) = \left[\frac{(0.7)(1.0)(0.87)}{\pi}\right](1.86 \times 10^{-2})$	$= +10.4°C$
$E/(h_c + h_r) = (-0.033)/(1.86 \times 10^{-2})$	$= -\ 1.8°C$
$(M_b/A_s)/(h_c + h_r) = (0.083)(1.86 \times 10^{-2})$	$= +\ 4.9°C$
$(T_c - T_a)$	$= +32.5°C$

[a] From Landsberg.[5]

with neither clothing nor sun, the difference $(T_c - T_a)$ would be reduced to only 8.3°C, or an environmental temperature of about 27.7°C. This last is, as it should be, quite close to the temperature of thermal neutrality noted just previously: 29°C.

Table 14-3
RELATIVE EFFECTS OF COAT THICKNESS AND COAT COLOR ON THE TEMPERATURE DIFFERENCE $(T_c - T_a)$

Coat thickness, L_c (cm)	0.0	0.5	1.0	2.0
$(M_b/A_s)R$ (°C)	2.4	10.7	19.0	35.6
Coat color, a_s	0.0	0.2	0.6	1.0
$a_s S/(h_c + h_r)$ (°C)	0.0	3.0	9.0	15.0

With the results of Table 14-3, it is possible to gain some understanding of the physical meaning of the role of clothing, which for other homeotherms is interpreted as the coat. The table contains values of the terms containing coat thickness and coat albedo, evaluated separately, in order to see the relative contributions of these factors to the difference $(T_c - T_a)$. One sees, for example, that for the man and the environment of Table 14-2 a coat of thickness 1 cm makes possible a contribution (19.0°C) greater than the maximum possible obtained by darkening the coat (15.0°C with $a = 1.0$). The extra "leverage" of coat thickness holds true for other, smaller homeotherms as well; so that a general aspect of evolutionary response to climatic change is suggested. It explains, for example, why the extra camouflage of white fur in the arctic is possible without great loss of insulation, despite its high albedo.

With a feeling for the relative magnitudes of the terms in Eq. (14-8), we may examine in broader, more general terms the role of clothing. Clothing worn in the arctic is loose-fitting, lightweight, and moisture-proof.[(4)] All these characteristics tend to maintain large positive values of the three factors on the right of Eq. (14-8). Loose clothing forms an insulating layer between the skin and the environment which is largely dependent on the trapped still air; a very large R. The looseness also permits withdrawing the arms into the clothing and folding them: smaller A_s. The light weight achieved through use of only one or two clothing shells conserves metabolic energy of motion and prevents the surface area A_x from becoming too large. Incidentally, though the factor A_x is absent from Eq. (14-8), it is there implicitly in the term L_c, the clothing thickness. Finally, the animal skins used in arctic clothing prevent E—always negative—from being too large. An added feature of Eskimo clothing is the hood, which extends well in front of the face and reduces the value of the convective transfer coefficient, h_c, for the energy-budget system.

The humid tropics pose the problem for clothing designers of keeping all positive terms in Eq. (14-8) small and all negative terms large, so that $(T_c - T_a)$ may be kept negative. Here the native clothing is as porous as possible—small R—without sacrificing the necessary protection against insect bites (Kelley[(4)]). It is light colored to reduce a_s—an almost irrelevant control in the arctic—and it is so constructed as to keep E large and negative by acting as a wick for perspiration. The wetness of the cloth also helps to reduce R, as seen in Table 14-1.

The climate of the desert is perhaps the most inhospitable of all. Not only does the environment change from hot to cold between day and night, but by day the resort to evaporative cooling is to be avoided because of the sheer lack of water to replenish perspiration. The principles applied in clothing design in the humid tropics are followed in the hot desert as well—loose fit, light weight, porosity, and light color. The necessary compromise between water shortage and the need for evaporative cooling is a difficult one, but clothing does reduce the magnitude of E under high operative temperatures, as seen in Fig. 14-5.

It should be said in concluding the discussion of clothing, that the principles and remarks made probably apply equally well in discussions of the roles of fur and other heat transfer adaptations in animals.

The Role of Housing

Housing may be thought of as an extension of the concept of clothing, and as such may be discussed in terms of Eq. (14-8). If the analogy is made, the processes taking place at the skin surface become those of the inner wall, and the shells of clothing become the walls of the building. The metabolic heat from the body core has added to it the heat from fuels burned for heating or cooking in the building, and vasomotor control has its analog in control over the movement of air within the rooms of the building. Although both clothing and housing shield their occupants against excesses in heat transfer and the violence of wind, precipitation, and lightning, it is with the problems of heat transfer we are most concerned here: the creation of what Landsberg[5] calls the "cryptoclimate."

As with clothing, housing in the arctic has the job of maintaining a large, positive $(T_c - T_a)$ in a continuously cold environment. The igloo is the typical response of man under these conditions, and it is a marvel of architecture developed through long periods of time using only the materials at hand.[6] The hemispherical shape of the structure presents the smallest possible A_s to the environment, at the same time enclosing the largest possible usable volume for a given amount of material. The shape is also the one most effectively heated from a small point source of radiant heat in the form of an oil lamp in the middle of the floor.

The wind-packed snow walls have a small value of k_c (see thermal conductivities in Table 4-2), and thus a large value of R for further maintaining a large, positive value of $(T_c - T_a)$. The interior heat soon forms a vapor-proof glaze in the interior walls of the igloo, and the entrance tunnel with its hanging skins is for all the world the same as the projecting hood of the parka. Finally, hanging skins from pegs on the inner walls may be seen as a further increase in R due to a reduction in k_b, as in contracted capillaries.

In the humid tropics, $(T_c - T_a)$ must be kept as negative as possible. The "parasol" housing, with sun- and water-shedding thatched roofs and virtually no walls, has low thermal resistance R, and free ventilation for increases in E. The roof provides the shade which reduces the value of S in the energy budget of the inhabitants beneath it.

In the desert, with its wide temperature contrasts between day and night, housing performs the jobs clothing cannot in the creation of a livable cryptoclimate. The thick mud walls not only provide a barrier against wind and dry air, they produce a cave-like temperature variation within. The temperature variations within a pueblo, for example, bear the same relationship to those in the air outside as the temperatures at depth in the soil bear to soil surface temperatures: greatly reduced amplitude and large time phase shift (see Fig. 4-3(b)). Thus, not only is the temperature within this structure of such materials moderate and comparatively constant, it reaches its maximum during the sleeping hours of smallest M_b and its minimum at the midday hours of greatest M_b. The mass the desert inhabitant cannot carry on his back to obtain this large thermal capacity, he builds into his house.

Although the variations of "primitive" architecture on the basic themes just described are many and sophisticated, present purposes need mention of only a few. In northern Nigeria, an outer layer of fiber matting provides a raincoat for the mud walls during the rainy season. Narrow, canyon-like streets and fountained courtyards open the otherwise massive and monolothic Mediterranean architecture to the needs of commerce. And in the classical tent is found the response to a need for what might be called a portable igloo.

What has civilized man, under high technology, added to the bag of architectural tricks? The change from the bearing wall to the curtain wall has enabled more and more use of glass, which, by virtue of the differential absorptivity shown in Table 14-4[(5)] permits both simple and complex forms of solar heating for housing. While man-made fibers and materials make possible very large values of thermal resistance, *R*, with very light weight, probably the key to today's architecture is the accurate and relatively

Table 14-4
TRANSMISSIVITY OF WINDOW GLASS IN THE SPECTRUM OF SOLAR RADIATION[a]

Wavelength (microns)	0.32	0.40	0.50	0.60	0.70	1.0	2.0	2.7	4.0	4.5
Transmissivity of glass	0	0.88	0.90	0.90	0.88	0.85	0.85	0.30	0.15	0

[a] Data from Landsberg.[(5)]

inexpensive internal controls over the cryptoclimate. Heating and air conditioning (cooling) make the architectural analogs of M_b large and small at will, thus exerting a tremendous leverage on the value of $(T_c - T_a)$. Requirements for these controls are often expressed analogously to the "varietal constants" for heat units in plants and animals. Thus, "heating degree days" are accumulated and form a measure of fuel requirements, by the formula (analogous to Eq. 11-1a):

$$\text{HDD} = \sum_N \left[T_t - \frac{(T_M + T_m)}{2} \right] \quad \text{when} \quad \frac{(T_M + T_m)}{2} < T_t \qquad \textbf{(14-9)}$$

where the threshold temperature is usually taken as 65°F (18°C). "Cooling degree days" are accumulated as a measure of air conditioning requirements by the formula in Eq. (11-1a), with a threshold temperature of 75°F (24°C).

Despite the environmental controls available to him, man's penchant for falling into inflexible habits—sentimentally or otherwise—has led him to build New England mansions on the northern Great Plains and thin-walled wooden houses with south-facing picture windows in Arizona. To use the phrase of Fitch and Branch,[(6)] "the worst (the modern architect) faces is a dissatisfied client. When the primitive architect errs, he faces a harsh and unforgiving Nature."

Whatever trouble man may have using his architectural tricks efficiently, he has used such schemes as those of the operative temperature (Eq. 14-5) to arrive at statements of what will constitute a comfortable environment for himself. Figure 14-6 shows, on a *TRe* diagram, combinations of temperature, relative humidity, and wind speed which will all produce an operative temperature of 69°F (20.5°C).[(5)] For calm air, this temperature is clearly within the standard winter comfort zone of the American Society of Heating and Ventilating Engineers, also shown.

The Limits of Tolerance

If the voluntary and involuntary controls on heat transfer available to an individual are insufficient to prevent departures of M_{Hp} from zero—either positive or negative changes in T_c—has he

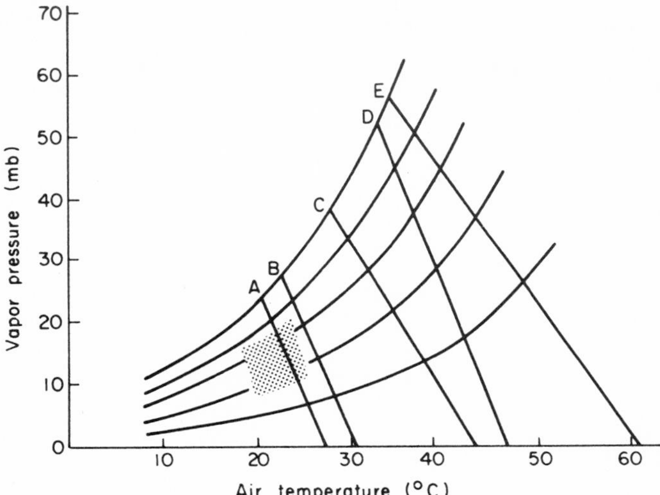

Fig. 14-6. The same operative temperature results from different combinations of air temperature, humidity, wind, and human activity. Curves A and B from Landsberg[5] for $T_{op} = 20.5°C$. Curves C, D and E from Herrington[1] for $T_{op} = 33°C$; these are upper tolerance limits. A: $T_{op} = 20.5°C$, $V = 0$; B: $T_{op} = 20.5°C$, $V = 1$ mps; C: $T_{op} = 33°C$, working nude; D: $T_{op} = 33°C$, $V = 0$; E: $T_{op} = 33°C$, V = 1 mps. Shaded area indicates ASHVE Standard Comfort Zone.

exceeded the limits of tolerance. Herrington[1] shows that the limits in a hot environment depend upon ventilation and upon the value of M_b produced by the individual's activity. Figure 14-6 also shows these limits on a *TRe* diagram. The similarity of the lines depicting the upper tolerable limits at about $T_{op} = 33°C$ and the lines depicting maximum comfort at about $T_{op} = 20.5°C$ is clear. The additional indication of the effects of an increasing M_b from work activity essentially completes the picture.

As for the cold tolerance limits of man, it should be remarked that by increasing M_b to perhaps 1.3 Met through high protein consumption, man seems to be able to maintain an even larger value of $(T_c - T_a)$ than otherwise[4]. There is a limit, of course, to man's metabolic response, and the gap between the 1.3 Met just noted and the 8 Met necessary to maintain $T_c = 36°C$ without clothing or shelter when $T_{op} = -1°C$ is simply unbridgeable.

Information cited by Herrington shows death has indeed been observed in experiments where the value of $M_{H\mathrm{p}}$ has exceeded 8 Met for an hour, producing a core temperature at death near $T_c = 22°\mathrm{C}$.[(1)]

Concluding Remarks

Despite the omission in this chapter of matters other than short-term responses to environmental stress, it should be clear there are enough connections between the plant, animal, and human branches of physiological ecology that researchers ignore these connections at the risk of fragmented, and thus less effective, progress. While we may recognize the complexity of the problem of relating atmospheric effects to biochemical and psychological effects in man, and certainly in animals, it may also be that progress in these areas will be much easier to come by if the matters discussed in this chapter are thoroughly understood. The absence of physical stress in the environment clearly does not insure the absence of mental strain, but it seems reasonable to suppose that it makes the absence of mental strain more likely.

REFERENCES

1. L. P. Herrington, "Biophysical adaptations of man under climatic stress," *Meteorol. Monograph* **8**, 30, Amer. Meteorol. Soc., Boston, Massachusetts, 1954.
2. See, for example, K. J. K. Buettner's "Physical aspects of human bioclimatology," p. 1112, *in* "The Compendium of Meteorology," Amer. Meteorol. Soc., Boston, Massachusetts, 1951.
3. D. H. K. Lee, *Intern. J. Biometeorol.*, **8** (1), 61 (1964).
4. See, for example, J. B. Kelley, "Heat, cold and clothing," *Sci. Amer.* p. 109 (February 1956).
5. H. Landsberg's "Bioclimatology of housing" *Meteorol. Monograph* **8**, 81, Amer. Meteorol. Soc., Boston, Massachusetts, 1954.
6. J. M. Fitch and D. P. Branch, "Primitive architecture and climate," *Sci. Amer.* p. 134 (December 1960).

PROBLEMS

14-1. For a sphere, the homeothermic energy requirement is proportional to the 2/3 power of the volume or the mass. Show that the the same relationship holds for (a) a cube and (b) a cylinder.

14-2. A cylindrical steel drum filled with a liquid simulates a large homeothermic animal. Its only source of heat is an electric element at the center of the drum simulating metabolic heat production, and the system has reached a thermal steady state where temperatures everywhere are constant with time. The drum has a diameter of 1 meter and a height of 2 meters. The heater element has a temperature of 40°C and dissipates 100 kcal/hour. If the average surface temperature of the drum is 20°C, what is the effective thermal conductivity of the model in cal/cm/min/deg?

14-3. If the cylinder model in Problem 14-2 is in an environment without sunlight, with a radiant (wall) temperature of 0°C, and an air temperature of 10°C:

(a) What is the change in operative temperature as the wind increases from 3.7 to 18.5 mph?

(b) What does the same change of wind speed do to the operative temperature if the cylinder has a 0.5 meter diameter?

14-4. Following is a temperature record for 10 days.

(a) How many Fahrenheit heating degree-days accumulated during the 10 days?

(b) On what day did 15 cooling degree days accumulate?

	Day No.									
	1	2	3	4	5	6	7	8	9	10
T_{max}	60	85	85	90	90	100	95	100	80	60
T_{min}	40	55	55	50	60	70	65	60	50	50

SECTION IV

The Urban Environment

Chapter 15

THE CLIMATE OF THE CITY

ALL OVER THE WORLD men are clustering together in towns and cities. As agriculture and commerce become better developed, the fraction of a population needed in rural areas declines, and the tendency of people to gather in urban commercial centers increases. Clearly, the urban environment is becoming of more direct concern to a greater proportion of humans every day in every part of the world. The necessity for knowledge of the urban environment, in all its aspects, to keep pace with the increasing urban population is among the most urgent of mankind's needs.

While it is true that energy exchange and interactions with weather are not nearly so well understood for systems like cities as they are for vegetation, the fundamentals are in hand and intensive research has begun. The intention in this chapter and the one following is to present several basic concepts and to make the reader aware of the general dimensions and the urgent importance of the problems discussed. The hope is that the reader will also see the relevance of the ideas and methods in previous chapters to some of the unexamined problems of urban climatology.

Basic Physical Urban–Rural Contrasts

Urban areas differ physically in five major respects from rural areas, the more so as the cities become "modern." The differences lie in (a) surface materials, (b) shapes of surfaces, (c) heat sources, (d) moisture sources, and (e) air quality. Cities are made up of more granite-like materials, with larger thermal capacities, than the countryside. Previous discussions of the soil heat budget make clear the importance of this difference. The three dimensional nature of the city, with its many buildings and multiple-level surfaces, present a complex geometry to the atmosphere. The city has major heat sources, such as industrial and domestic heating and automobiles, not found in such concentration in the country. The rural areas, on the other hand, are sources of moisture relative to the city. The comparative absence of vegetation and the efficient removal of precipitation in the city assures this contrast. Finally, the wastes of industry, domestic activity, and modern transportation all contribute to major changes in the quantity and quality of materials suspended in the urban atmosphere.

In this chapter, the discussion will be built on consideration of the climatic variable which is in many ways the key to urban–rural contrasts: temperature. The relevance of the five kinds of difference just listed, and the nature of the differences in other climatic variables will be developed as well.

Urban–Rural Temperature Contrasts: Theory

As in previous chapters, it will be fruitful to obtain a form of energy-budget equation on which to base discussions of urban–rural temperature contrasts. To begin, we consider both the city and the nearby countryside to form Type 3 energy-budget systems and proceed to write appropriate equations describing unit areas of the two environments for a unit of time. First, for the city, an equation similar to Eq. (7-20) but employing mathematical descriptions of radiation and convection developed in Chapter 14 is

$$a_s S_{di} + \varepsilon\sigma(T_r^4 - T_U^4) + h_c(T_a - T_U) + E + B + M + H_h + E_h = 0, \quad \textbf{(15-1)}$$

where T_U is the effective temperature of the urban "canopy" of buildings, towers, trees, etc.; T_r is the effective radiative temperature of the sky; and T_a is the temperature of the free air just above the canopy. Temperature measurements within and above cities have shown this canopy depth to be only two or three times the height of a typical building in the city. For a rural canopy with effective temperature T_R, the analogous equation is

$$a_s'S_{di}' + \varepsilon'\sigma(T_r^4 - T_R^4) + h_c'(T_a - T_R) + E' + B' + M' + H_h' + E_h' = 0. \tag{15-2}$$

If, as in other discussions, we let $T_r = T_a$, and $\varepsilon = \varepsilon'$, the considerations of radiation and convection are more manageable. It will also serve well to let $S_{di} = S'_{di}$, saying that these represent the shortwave energy arriving at the top of each "canopy." The differential effects of reflective geometry and of the increased albedo of dirty air are then contained in the differences between a and a_s'. Finally, if the general windflow is across the city–country boundary, and the advection from the country equals, the advection to the city, we say $H_h + H_h' = 0$ and $E_h + E_h' = 0$. With these assumptions, subtract (15-2) from (15-1) to get

$$(a_s - a_s')S_{di} + \varepsilon\sigma(T_R^4 - T_U^4) + h_c(T_a - T_U) - h_c'(T_a - T_R) + (E - E') + (B - B') + (M - M') + 2(H_h + E_h) = 0. \tag{15-3}$$

Making the approximation leading to h_r (see footnote on page 246), and letting $h_c' = h_c + \Delta h$, (15-3) becomes, upon solving for the urban–rural temperature difference,

$$(T_U - T_R) = \frac{(a_s - a_s')S_{di} + \Delta h(T_R - T_a) + \Delta E + \Delta B + \Delta M + 2(H_h + E_h)}{(h_r + h_c)}. \tag{15-4}$$

In Table 15-1 may be seen the impacts of various changes in the terms of Eq. (15-4). Probable algebraic signs and illustrative numbers for the various terms demonstrate the expected behavior

Table 15-1

RELATIVE NUMBERS ILLUSTRATING THE CONTRIBUTIONS OF VARIOUS FACTORS IN EQ. (15-4), UNDER DIFFERENT COMBINATIONS OF AIR QUALITY, WIND DIRECTION, AND TIME OF DAY

Conditions	$(a_s - a_s')$	$(a_s - a_s')S$	$(T_R - T_a)$	$\Delta h(T_R - T_a)$	ΔE	ΔB	ΔM_{Hp}	ΔM_b	$(H_h + E_h)$		$(T_U - T_R)$
Daytime: clean air, wind across boundary	+	+2	+	0?	+1	−1	−1	+1	−1	=	+1
Nighttime: clean air, wind calm		0	−	0?	0	+1	+1	+2	0	=	+4
Daytime: dirty air, wind across boundary	+	+1	+	0?	+1	−1	−1	+1	−1	=	0
Nighttime: dirty air, wind calm	Relative longwave flux +1		−	0?	0	+1	+1	+2	0	=	+5
Daytime: dirty air, wind parallel to boundary	+	+1	+	0?	+1	−1	−1	+1	0	=	+1

Notes: a_s is always +; S always +, but $S_{dl} < S_{dl}'$ when city air is dirty.
The signs of $(T_R - T_a)$ indicate lapse by day and inversion at night.
The signs and magnitudes of Δh are among the objects of current research in urban climatology.
Daytime evapotranspiration in the country is large and −, hence ΔE is +.
Daytime soil storage is − and is largest in the city, hence ΔB is −.
The same remarks apply to storage in the canopies.
Heat production by combustion is always +, but greater at night.
When the general wind is along the urban–rural boundary $H_h = H_h'$ and $E_h = E_h'$.

of the temperature contrast under several sets of important conditions. The notes beneath the table will explain the signs in the entries of the table. The justification for assigning a zero value to the differential convection coefficient, Δh, after saying in the notes its values and signs are unknown, is just this: Whatever increase in h_c' may accrue because of greater wind speeds in the country is likely to be offset by the greater amount of heat exchanging surface within the canopy of the city. To repeat, the numbers in Table 15-1 are only illustrative in the absence of generally available and acceptable values in the literature of urban climatology.

Examination of Table 15-1 reveals many interesting results. First, for a city with clean air and with wind flowing from the country into the city, the urban maximum temperature exceeds the rural maximum by a small amount. Under the same air quality conditions, the relative warmth of the city increases at night. This increase would be present even if the wind were not calm, and it is due primarily to the increased fuel consumption and the recovery of heat stored by day in streets and buildings.

A second comparison shows that when the urban air is dirty and reflects more sunlight from airborne contaminants, the temperature contrast is reduced during the day, as compared with the case having clean air. At night, on the other hand, the temperature contrast is increased because of the partial closing of the radiative "window" to the exit of longwave energy from the city's canopy.

Finally, it becomes clear that when cooler air is not being advected into the city, the temperature contrast rises. Thus, in a situation where the urban area has a long and linear shape, such as the "megalopolis" lying along the northeastern seaboard of the United States, a general windflow along, rather than across, the upwind urban boundary will tend to increase the temperature difference, $(T_U - T_R)$.

Urban–Rural Temperature Contrasts: Observations

How do observations compare with the theoretical description of urban-rural temperature contrasts? Observers have known for some time that such differences exist,[(1)] but the spatial and temporal details of the contrasts are still being explored. Much of the problem in deducing the true city climate from meteorological observations lies in the fact that the observation sites are too seldom comparable. As Landsberg puts it, "when authors compared data obtained from roof stations (T_U) with others where the instruments were exposed on the ground (T_R), they likely masked the very information they searched for." Other problems of comparability exist, but for present purposes it is enough to say caution is necessary.

Figure 15-1, from Mitchell,[2] shows diurnal temperature traces from a pair of stations at Vienna for two contrasting

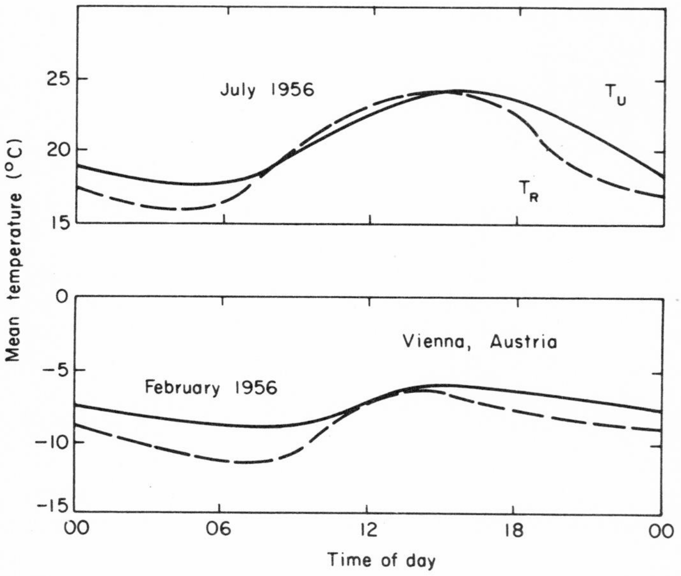

Fig. 15-1. Representative warm-season and cold-season diurnal temperature traces for a pair of urban and rural stations in Vienna, Austria. From Mitchell.[2]

seasons. Urban climatologists agree these traces are typical of large midlatitude cities everywhere. Clearly, the temperature contrast is greater at night, as suggested by the contents of Table 15-1. In addition, what is not shown in Table 15-1 is the fact that the city record exhibits even greater diurnal asymmetry and delay in temperature maximum than does the country. The discussion of asymmetry and "thermal inertia" in connection with Fig. 4-1 makes the urban record quite reasonable, but what of the spatial patterns of temperature difference?

Urban–Rural Temperature Contrasts: The "Heat Island"

It is now quite well established that isotherms in the vicinity of any town or city tend to show a pattern like the one for London in Fig. 15-2, where temperatures are greatest in the most densely

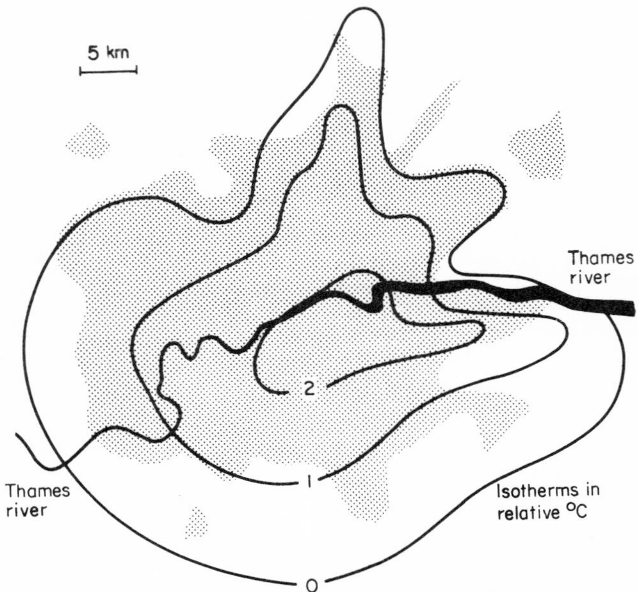

Fig. 15-2. Representative isotherms (in relative °C) illustrating the "heat island" effect of a city. After Chandler's study of London.[3]

built up portions. The appearance of the isothermal pattern has suggested the name "urban heat island" for the pattern and its several variants.

Mitchell[2] makes the case very nicely that, while in some cities topographical effects on climate may contribute to an apparent heat island, the island can be shown to be man-made—the direct result of the existence of the city itself. Three kinds of evidence support Mitchell. First, in areas of topographic simplicity, such as London, the heat island shows up very clearly. Second, the intensity of the contrast—the "height" of the island—increases as the city's population grows. Third, analysis of data on $(T_U - T_R)$ for weekends and work days separately shows clearly that the increased values of M_b on work days in the heart of the city produce greater contrasts on those days. Figure 15-3 shows Mitchell's data relating rates of population growth and relative urban warming,* while Table 15-2 gives his results on weekend–workday differences for New Haven, Connecticut. There is qualitative agreement between Tables 15-1 and 15-2.

Table 15-2

URBAN-RURAL CONTRASTS IN MAXIMUM AND MINIMUM TEMPERATURE, AS A FUNCTION OF DAY OF THE WEEK[a]

	$(T_U - T_R)$ (°F)		
Day of the week	Maximum temperature	Minimum temperature	Mean temperature
Sunday	0.1	1.2	0.6
Monday–Friday	0	2.2	1.1
Saturday	0	2.1	1.0

[a] Data from Mitchell.[2]

* Mitchell's argument for using the change in the square root of the population as a measure of growth is roughly as follows. Assuming a constant population density, D, throughout a city, and assuming a roughly circular area, A, having radius r, these variables are related according to

$$\text{Population} = DA = \pi r^2 D; \qquad \text{or} \qquad r = (P/\pi D)^{1/2}.$$

If the amount of warming is considered proportional to the travel distance of air moving from the urban–rural boundary to the heart of the city, r approximates this distance and $P^{1/2}$ is the correct measure for this purpose.

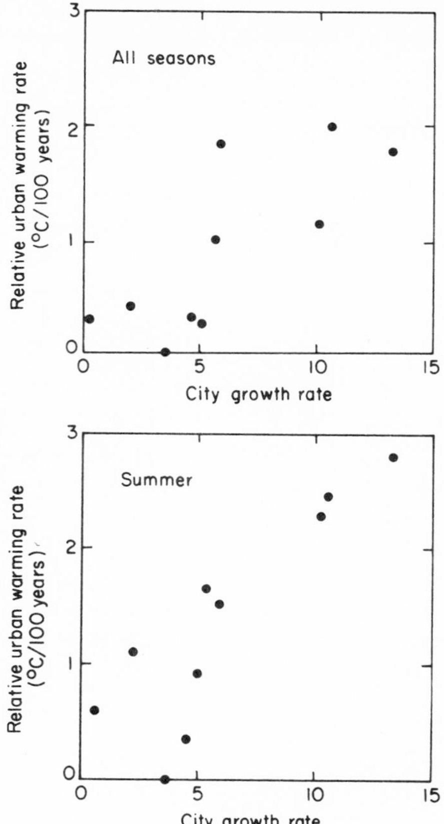

Fig. 15-3. Intensification of the "heat island" effect as a function of growth rate of the urban area involved. After Mitchell.[2]

At various points in the discussion it has been suggested that the intensity of the heat island, as indicated by the value of $(T_U - T_R)$, depends upon topographic setting, population, density of population, and map geometry. It is clear the intensity must depend upon such other things as the relationships of land and large water bodies near the city, the latitude as it affects the solar energy supply, and the regional weather conditions. Sundborg[4] showed some time ago that a good empirical relationship connecting the nocturnal urban–rural temperature contrast and regional weather is

$$(T_U - T_R) = (a - bN)/V \qquad \textbf{(15-5)}$$

where N is the number of tenths of cloudiness ($0 \leq N \leq 10$), V is the regional wind speed, and a and b are constants for the locality in question. The point of the equation is that the contrast is heightened most on nights with clear skies and calm winds, as might indeed be expected from the discussions in Section 1 of this book.

In summary, then, it is the magnitude of the physical contrasts between city and country, as affected by several variables, which determines the climatic contrasts. A very populous city may have such a suburban "sprawl" that it blends very gradually into its rural surroundings and the horizontal temperature gradients become very small. Thus it has been shown that even a small city exhibits a distinct heat island if it contrasts physically with its immediate environs.(5) Finally, as to the effects of regional weather for temperature contrast, it matters not what the weather map looks like, so long as the requisite conditions of clear skies and calm winds are met locally.(5)

Other Urban–Rural Contrasts

In Table 15-3 are the representative values of urban–rural difference in a number of meteorological variables, as given by Landsberg,(6) after exhaustive literature review. Temperature differences have already been discussed at length, and reductions of solar energy and moisture in the city have already been noted. The effects of the city's air quality on the very short-wave portion of the solar spectrum—the ultraviolet—are striking. It should be added that the city exhibits a deficit in absolute, as well as relative, humidity on the average; thus temperature alone does not account for all the difference in relative humidity given in the table. It is easy to imagine that the frictional braking of the wind by the

Table 15-3
REPRESENTATIVE VALUES OF URBAN-RURAL CONTRAST FOR SEVERAL WEATHER VARIABLES[a]

Variable	Urban compared with rural
Annual mean temperature	0.7°C more
Mean winter minimum temperature	1.4°C more
Solar radiation on horizontal surfaces	15% less
Ultraviolet radiation	15% less summer; 30% less winter
Annual mean relative humidity	6% less
Seasonal mean relative humidity	8% less summer; 2% less winter
Annual mean wind speed	25% less
Speed of extreme wind gusts	15% less
Frequency of calms	15% more
Cloudiness frequency and amount	10% more
Frequency of fog	30% more summer; 100% more winter
Total annual precipitation	10% more
Days with less than 0.2 in (5 mm) precipitation	10% more

[a] From Landsberg.[6]

buildings and towers of the city can amount to 25%. Not shown in the table, but mentioned by Landsberg[1] are two striking consequences of the temperature differences: 14% decrease in the number of precipitation days with snow, and typically three-week increases in the length of the freeze-free season in the city.

What of urban effects on visibility or visual range? In Table 15-4 Landsberg's figures show clearly the urban effects on visibility reduction due to fog, but comparable results on reduction due to particulate contaminants in the air are less readily available. Landsberg claims his figures for Detroit are representative: the increase in annual number of hours having reduction to less than 1 mile due to "smoke" is from 5 hours in the country to 49 hours in the city.

Table 15-4 provides a transition to the subject of cloud and

Table 15-4
URBAN–SUBURBAN–RURAL CONTRASTS IN VISUAL RANGE DUE TO OCCURRENCE OF FOG.[a]

Type of visibility reduction due to fog	Summer			Winter		
	City	Suburb	Country	City	Suburb	Country
Light fog (¼ to 1 mile)	49[b]	49	6	350	219	60
Moderate (¼ mile to 300 ft.)	3	3	2	49	43	28
Dense (less than 300 ft.)	0	1	0	8	14	5

[a] Data are for Paris, France, from Landsberg.(1)

[b] These data are for 9 A.M. observations. expressed as days per 1000 days.

precipitation differences between country and city. The figures on these variables in Table 15-3 need some explanation in light of such apparent contradictions as an increase in precipitation and a reduction in humidity in the city. Figure 15-4, showing an idealized model of the airflow, polluted layer, and cloudiness near a city, will help to explain these differences.

The tendency shown for air to rise over the city center is

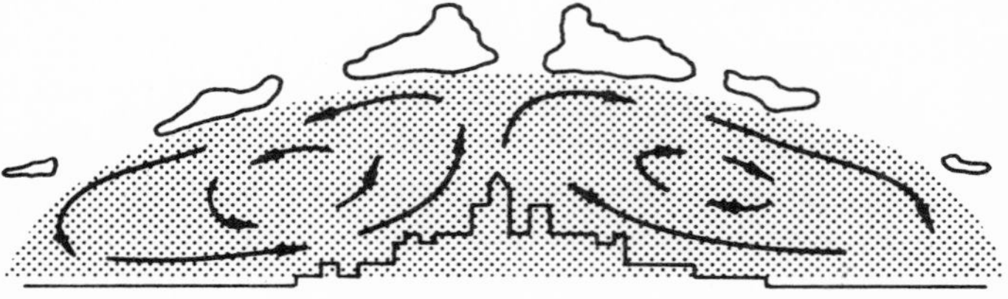

Fig. 15-4. Schematic of the local air circulation, differential cloudiness, and shape of the "haze hood" above an idealized city. After Lowry.(7)

offered as explanation of part of the increase in both cloudiness and precipitation amount, since this rising motion is a fundamental requirement for any kind of convectively produced precipitation. The tendency for air to subside in completing the flow system over the countryside, of course, also works to accentuate the urban–rural contrasts in cloud and precipitation. Such a circulation as shown in Fig. 15-4 would be dissipated completely under conditions of strong regional windflow. The conclusion is reached, therefore, that the majority of the increases in cloud and precipitation on an annual basis come as the result of small increments on relatively calm, moist days when the convection over the city is well developed. This conclusion is borne out by the last entry in Table 15-3, pertaining to days with light drizzly rainfall.

As for clouds and precipitation associated with fog, the explanation is more complex. Table 15-4 shows that the great majority of foggy days are in the winter. They are probably initiated by arrival of a fresh, cold air mass of regional dimensions, which results in an increase in urban heating by combustion of fuels. The burning produces both water wapor and particulate contaminants, the particles becoming condensation nuclei for the vapor as the cold city begins slowly to warm up. Until the warming becomes substantial, however, the condensed excess of vapor lingers as a fog. The increased albedo of the top of the foggy layer, of course, will retard the warming, and the fog becomes, like a blanket of snow, self-perpetuating.

Advection fog may form when cold, moist air flows gently into the city from the countryside. In the suburbs, the cold air encounters the city's oversupply of condensation nuclei before it has been warmed very much. The result is fog in the suburbs at times when fog is absent in both the country and the city center. This process may account for the greater number of suburban fog days in all categories of Table 15-4 than one would expect from an interpolation between city and country.

It should be said in closing that the apparent contradiction of more cloud and precipitation and less moisture in the city, on the average, is explained simply by noting that it is on the average.

The majority of days are "clear and dry" and overbalance the "cloudy and moist" days in an annual mean value. The facts remain, however, that there is in every sense a true urban climate which would not be in existence were it not for the presence of the city. As noted in the beginning of this chapter, this increasingly important example of man's modification of an environment is overlooked only at great peril. Every student of biometeorology, or of ecology for that matter, ought to be aware of this and other examples of the large-scale, inadvertent modification of climate by the hand of man.

REFERENCES

1. H. E. Landsberg, The climate of towns, in "Man's Role in Changing the Face of the Earth," p. 584, Univ. of Chicago Press, Chicago, Illinois, 1956.
2. J. M. Mitchell, "The thermal climate of cities," p. 131 in Air over Cities, U.S. Public Health Service (1962); also in *Weatherwise* **14** (6), 224 (December 1961).
3. T. J. Chandler, Biometeorology 2, *Proc. 3rd Intern. Biometeorol. Cong.*, p. 589, Pergamon Press, New York, 1967.
4. A. Sundborg, *Tellus* **2** (3), 222, 1959.
5. R. Hutcheon *et al.*, *Bull. Amer. Meteorol. Soc.* **48** (1), 7 (1967).
6. H. E. Landsberg, City Air—better or worse?, in "Air over Cities," p. 1, U.S. Public Health Service (1962).
7. W. P. Lowry, *Sci. Amer.* p. 15 (August 1967).

PROBLEMS

15-1. Examine the effects of urban surface geometry on absorption of solar radiation in the following way. Consider 2 cubes, of material with albedo 0.5, with edges oriented north-south and east-west, one cube directly north of the other. As functions of elevation angle of the sun and of relative spacing between the cubes, estimate:

(a) The shortwave energy flux (relative to the direct solar beam) on the southernmost south-facing wall,

(b) the relative shortwave energy flux on the north-facing wall between the cubes, and

(c) the mean albedo of the whole "city."

15-2. Examine the effects of surface materials on urban–rural temperature contrast as follows. As the rural surface, use a horizontal plane of wet sand; and as the urban surface use a horizontal plane of granite (Table 4-2). Now estimate the amount of temperature rise of a centimeter depth of each material between 8 A.M. and noon if the energy flux on a horizontal surface is that marked R_n in Fig. 8-1. Assume the entire B term is used to heat only the surface centimeter, and that the energy distribution is

	B/R_n	H/R_n	E/R_n
Wet sand	0.2	0.3	0.5
Granite	0.4	0.6	0.0

15-3. Discuss likely reasons for the scatter of points in Fig. 15-3.

15-4. A city is generally about 5°C warmer than the countryside on a night with no clouds and a light wind of 0.5 mph. The temperature excess is only about 0.1°C on a night with complete overcast and a wind of 10 mph. What temperature difference would you expect on a night with:
(a) half the sky cloudy and a wind of 5 mph?
(b) a complete overcast and a wind of 1 mph?

Chapter 16

AIR POLLUTION METEOROLOGY

DISCUSSION OF THE QUALITY of the air in the urban environment, and of the effects of contaminants on climate in and around a city, necessitate a more detailed inspection of the sources of the contaminants and the manner in which the contaminants become a part of the atmosphere.

From Chapters 4 and 6 the reader will recall the basic connections among the concepts of atmospheric stability (as measured by the temperature lapse rate), turbulence, and turbulent transfer. In essence, we find that turbulent eddy motion and the resulting transfer processes are more vigorous in the presence of strong lapse conditions—large temperature decrease with height—and are suppressed in the presence of inversion conditions. The turbulent motion is made up of essentially random mass transfer, the eddies being thought of as superimposed upon an underlying uniform flow.

Several notions must now be added to those just mentioned for the foundation of the chapter to be complete. First, it must be said explicitly that there are eddy components acting in all directions, not just the vertical. This was implied in the footnote to Eq. (8-5), but it must now be emphasized. Second, it must also be said explicitly that random motion may exist in the absence of any underlying flow; that is, when $U = 0$ in the sense of Chapter 6. The image of an expanding sphere moving out from a central point was used in the discussion of thermal diffusivity in Chapter 4. Common experience tells us that mass, in the form of smoke particles, also moves outward from a common point after an event such as an explosion. Thus, the *diffusion* may take place equally in all directions, or it may be superimposed on underlying flow. Finally, it is useful to think of the emissions from such a source as an industrial chimney, or "stack," as a sequence of very closely spaced puffs. We may now examine in rather small scale, and then on a larger scale, the behavior of these effluents after they leave their source.

Diffusion of Contaminants: The Microscale

Figure 16-1 shows three things. First, it shows an instantaneous view—a "snapshot"—of a plume from a stack. Second, it shows

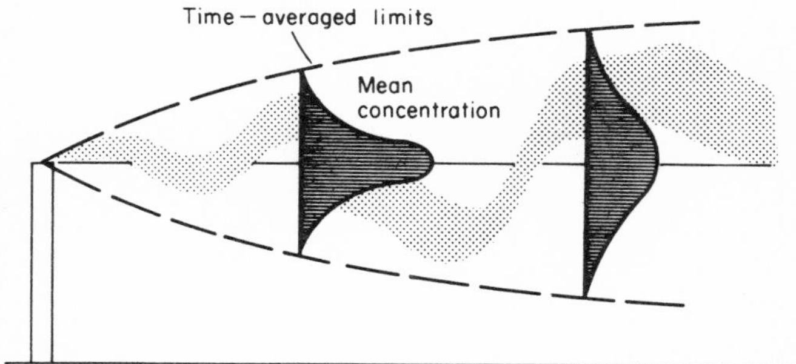

Fig. 16-1. Schematic of the relationships among the momentary appearance of a plume, its time-averaged concentration of contaminants downstream, and the time-averaged limits of its vertical envelope.

the outline, or envelope, within which the plume is confined over a certain period of time—a "time exposure." Third, the figure shows the kind of variability of the mean concentration of contaminant one may expect to find within the envelope of the plume. The highest mean concentrations lie along the axis of the plume, and the uniformity of mean concentration in any cross section of the plume increases downstream. It should be noted that any given portion of the plume in the snapshot did not follow the shape of the plume in travel from the source to its present position. Rather, the lower portions likely left the stack at a moment of downdraft, the upper portions at a moment of updraft, and all followed much more linear trajectories than the one suggested by the momentary shape of the plume.

As suggested above, atmospheric stability has a great deal to do with the geometry of a plume. Figure 16-2 depicts five basic

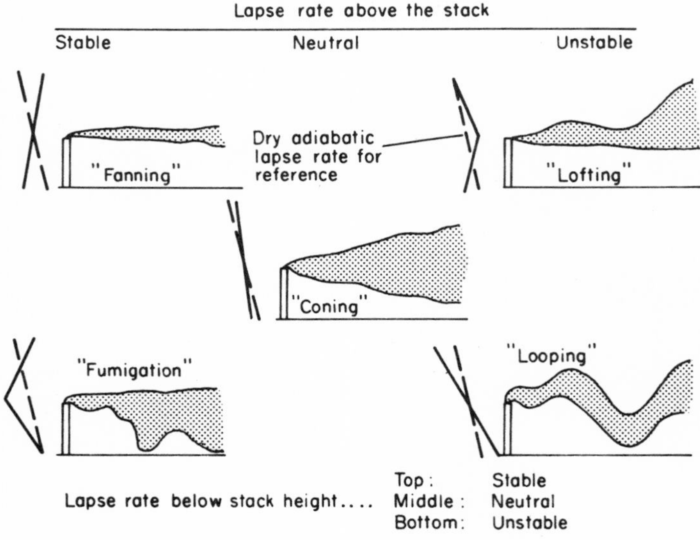

Fig. 16-2. Names and appearances of plume types related to stability of the lower atmosphere below and just above the effective height of the source stack.

types of plume and the lapse rate conditions producing them. When vertical and horizontal motion are suppressed both below and above the stack height, the plume spreads only slightly downstream. Such a plume, which spreads more laterally than vertically, takes its name from the fan-like appearance seen from above; and it may be traced intact many kilometers downwind from the source. When lapse conditions exist both below and above the stack height, the plume "loops" and gives a very disorganized appearance. Intermediate conditions of stability produce a cone-shaped plume, while lapse conditions above a ground-level inversion to stack height result in upward dispersion known as "lofting." Finally, the potentially most dangerous and perhaps least understood plume is the type which results when morning heating of the surface layers—much as suggested in Fig. 4-8a—finally reaches the height of a fanning plume and suddenly brings high concentrations of pollutants downwards to the surface everywhere beneath the plume axis: "fumigation."

Mathematical models exist with which one may estimate the mean concentrations at any point in space downstream from such continuous point sources as have been described. In addition to the information in Fig. 16-1, the models produce estimates such as those in Fig. 16-3, which should give the reader a better appreciation for the interaction of the plume and the underlying surface.

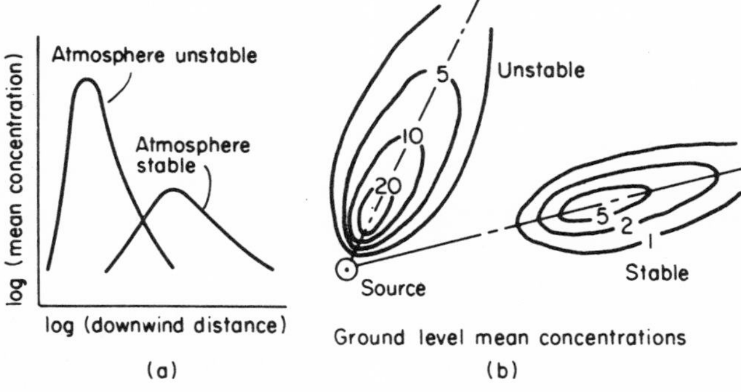

Fig. 16-3. (a) Typical predictions of time-averaged pollutant concentration at ground level beneath the centerline of a plume from a single point source, as a function of atmospheric stability. (b) Typical predictions of ground-level patterns of average concentration beneath plumes in stable and unstable airflow.

In Fig. 16-3a the mean concentrations at ground level beneath the plume axis, or centerline, are shown as a function of downwind distance. This information is extended to the crosswind distance in Fig. 16-3b. Clearly, maximum ground level concentrations are found some distance downwind if the source is elevated as in Fig. 16-1. What is more, the downwind distance to the maximum increases as the atmospheric stability increases, as suggested in the various parts of Fig. 16-2.

Though the mathematical models and their results give the appearance of considerable precision in describing the time-averaged behavior of a plume, in practice the accuracy is only moderately good. Several things produce departures from what the models predict, such as the problem of estimating the true diffusing power of the turbulent field from wind measurements, and the length of time over which the averaging is made. For example, the continuous record of an air sampler beneath the axis of a looping plume would probably show occasional sharp rises to high concentrations, interrupting an otherwise continuously low value. If a short-term average from this record included several of these "bursts," the time mean would be considerably higher than a longer-term average which included more of the quiet time between bursts. If the averaging time is too long, however, there is a greater chance the direction of the mean wind will have changed and the model's accuracy further disturbed.

The mention of bursts of high concentration at ground level beneath a looping plume further requires mention of the concept of dosage. For some biological reactions to pollutants a short exposure to a given amount, or "dose," of the chemical may be as severe as the reaction to the same amount spread over a longer time. Thus two forms of apparent contradiction may be explained by noting the concept of dosage. First, damage to organisms may occur under conditions of a very low mean concentration averaged over a relatively long time. Second, under conditions of great tendency for diffusion of pollutants—not at all the conditions generally associated with an air pollution problem—great damage to organisms may occur in the course of a short-lived burst of high concentration, as beneath a looping plume.

Another factor which reduces the accuracy of predictions is the rise of the plume above the height of the stack's orifice, which requires an adjustment in the model's value of the height of the plume axis. This "plume rise," shown in Fig. 16-4, is due in part to the momentum which the mass of effluent has as it leaves the orifice and in part to the buoyancy it has if it is warmer than the environmental air. Methods exist for estimating the plume rise, so that on balance the mathematical models for predicting mean

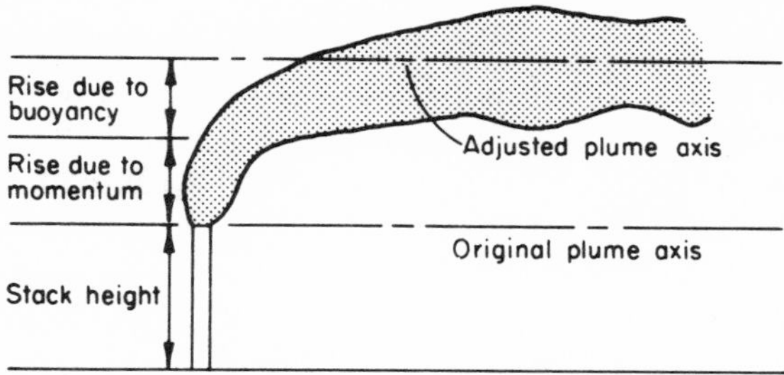

Fig. 16-4. Schematic of a single plume showing the adjustment necessary in height of the centerline because of momentum and buoyancy of the plume.

concentrations are still quite useful, although not of greatest accuracy. A good deal of their usefulness comes from the assistance they give in discussions such as those here.

Diffusion of Contaminants: The Mesoscale

By the time pollutants move well away from their sources, and plumes become indistinguishable from one another, discussion of air pollution meteorology must be undertaken from the point of view of the mesoscale. To begin, recognize that the atmosphere acts upon the contaminants in two basic ways: It disperses them and it dilutes them. Dispersion and dilution are not entirely separate effects, but there are differences between them which are important. Dispersion implies the transport of contaminants to places other than their point of entry into the atmosphere. Dilution implies the reduction of concentration by thorough mixing with large volumes of "clean" air.

If a puff, or a continuous flow, of contaminants is transported intact downstream, as in a fanning plume, there is dispersion without dilution. If the puff, or continuous stream, is mixed with large volumes of air in the neighborhood of its source, there is dilution without dispersion. The fact is that dispersion and dilution usually occur together, since moving air becomes turbulent and mixes the contaminants throughout its entire volume.

To extend consideration of dispersion and dilution to further understanding of problems of air pollution meteorology, consider a simple model of the atmosphere, as shown in Fig. 16-5. First, there are two layers in the model, one above the other. The depth of the lower layer is determined by nearby topography and by the sensible heat transfer rate at the surface, which heating results in

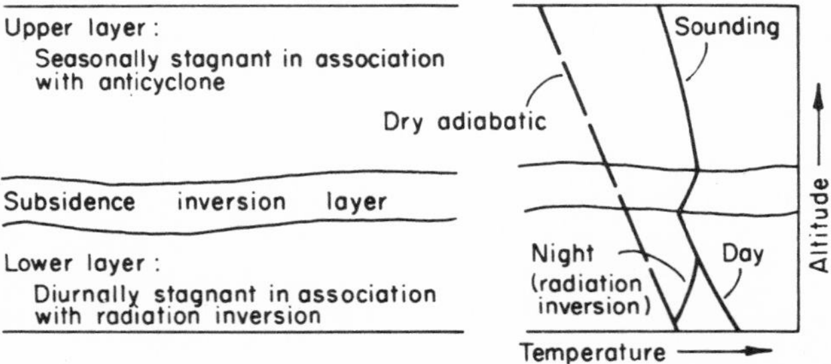

Fig. 16-5. Schematic model relating diurnal and seasonal potential of the atmosphere for dispersion of air pollutants on the mesoscale.

the vertical mixing of thermal turbulence. The depth of this lower layer varies from place to place and from one time to another, but is usually something like 1 km. The upper layer extends from there essentially on up to 10 km.

The second feature of the model is that both layers may be in motion or stagnant, to some extent independently of one another. As a general rule, the upper layer is stagnant or in motion on a seasonal basis, while the stagnation and motion of the lower layer comes and goes on a daily basis with the appearance and disappearance of the surface inversions of Fig. 4-8.

Third, these two layers may be either coupled (i.e., connected) or uncoupled from one another, a characteristic which in general changes seasonally. The uncoupling occurs when a layer of air exhibiting a negative lapse rate appears between the two layers, as shown in the temperature profile of Fig. 16-5. From previous discussions we know vertical motion is suppressed in such a layer which thereby presents a barrier to the sharing of any horizontal motion between the two layers. Previous discussion has explained the formation and disappearance of the inversions in the lower layer, but what is the explanation for the decoupling inversions between the layers?

Although most of the explanation must be left to standard textbooks on meteorology, it may be said here that the decoupling inversions result from macroscale sinking of air from great heights —usually from near the top of the upper layer—in connection with features called "high pressure areas" or "anticyclones." The air in sinking is heated by compression, and if it is prevented from subsiding all the way to the earth's surface by the presence there of a cooler, more dense mass of air, the stratification in the decoupled model results. In areas where the surface air mass is renewed daily by a cool inflow of air from the ocean, as in most midlatitude west-coast locations, the decoupling persists through many weeks.

Equipped with an understanding of this simple model, one may appreciate the varying degrees of potential for accumulation of pollutants near the earth's surface. The least serious combination of conditions occurs when both layers are in motion and are coupled to one another. Under these conditions both dilution and dispersion are excellent, and accumulation of pollutants is no problem. Intermediate in seriousness are such circumstances as those which arise during the season when the upper layer is stagnant and decoupled from the lower layer, but the lower layer is in motion most of the day. Here pollutants may be moderately well diluted, but in recycling on the mesoscale in a circulation such as the various local wind systems described earlier—the land-sea breeze or the mountain-valley wind—they are not really dispersed.

The most serious of all conditions, promoting accumulation of constantly emitted pollutants, occurs when both upper and lower layers are stagnant. Under such circumstances even the diurnal tendency for motion in a moist lower layer is suppressed as nocturnal fog retards surface warming during the day. It is these occasions which are described in relation to fog frequency in the last chapter and which have accompanied the worst air pollution disasters to be documented to date.

Concluding Remarks

Although many aspects of air pollution meteorology have been omitted in this discussion, the reader should have a feeling for the complexity of the subject. The problems are many times more complicated than the often-heard comments to the effect that only in times of inversion does an air pollution problem exist. On the one hand, an inversion as such is not necessary for suppression of motion and stagnation, since even a tendency toward an isothermal temperature profile will reduce the diffusive power of the atmosphere. And on the other hand, it is obvious that stagnation does not present a problem in the absence of dangerous pollutants. Unfortunately, progress in air pollution meteorology, and in most mesoscale problems of meteorology, is retarded by the intermediacy of the size scale involved. Too large to be observable from weather stations and instrumented towers and too small to be detected by the widely spaced stations of the regular networks maintained by governments around the world, the processes at work here must be observed by specialized and very expensive means employed for just this purpose. Thus, observations are accumulating at a much slower rate than the need to utilize them in research.

This same scarcity of data on reactions of organisms to air pollution in part explains the absence here of any discussion of receptor effects. Such wide areas of investigation as the modification of pollutants during transit in the atmosphere, the complexities of synergistic effects when a combination of pollutants does damage to tissue none of the pollutants would do if present alone, and the effects on life of ionized molecules in the atmosphere—all these are omitted in favor of a discussion of the meteorological processes underlying the increasingly serious problem of air pollution.

Chapter 17

A LOOK BACKWARD AND A LOOK FORWARD

No doubt the remarks to follow would be as applicable to other scientific fields in a similar stage of development as they are to biometeorology today, but few scientific fields are surrounded with more urgency of need for the future well being of life as we know it. It is clear to ecologists today that the rate of man's modification of the environment of earth is so very much greater than either the rate of increase of his knowledge about the environment or the possible rate of healthy natural adjustment on the part of the environment, that either the rate of modification or the rate of increase of knowledge must change drastically to keep things within bounds.[(1)]

Many of the ideas of biometeorology are very old indeed. The most skillful hunters and the most efficient gatherers among ancient men knew many of the basic concepts described in this book. The successful early farmers learned many more of these concepts, and all passed them on to their successors. By trial and error the knowledge came slowly and was hard-won.

The really rapid growth in what we now call biometeorology has come with the advent of routine weather observations, of statistical methods, and of high speed computers. The routine observations have provided the raw materials of research, which has in turn led, by means of clues disclosed by statistical analyses, to insight and understanding about general principles of organism–environment interaction. Computers have permitted the sifting from much more data of clues about even more complex interactions, and have enabled the great strides to understanding possible through mathematical simulation of complex systems of variables.

Today we speak in great detail of some processes, such as energy budgets of simple systems and some aspects of photosynthesis. But in fact our understanding overall extends to only a rudimentary awareness of the broad perspectives of weather and life. We sense the deep seated implications of manipulation without complete knowledge, but even with the availability of fairly sophisticated methods we are only on the verge of understanding the true nature of the totality of organismic requirements for survival.[(2)] Such easily conceived but intractable ideas as "trigger reactions," in which the environment begins an irreversible biological process subsequently insensitive to the environment, and "delayed reactions," in which the manifestation of an environmental effect appears in the organism only after considerable delay, are still beyond our reach. The social interactions among organisms, when added to the biophysical interactions, make the ecologist's task seem nearly impossible.

But the task is not impossible. Even now our understanding has pointed the way to improvements which have not been put into practice. We know the usefulness of, but have not yet begun the collection of, routine observations of data on solar radiation, soil temperature, and phenological events, to name a few. Our knowledge, in short, is far ahead of its application. The increasing realization that the subject of ecology is a vital one is itself a hopeful sign. To repeat the phrase offered earlier, the task is to keep things within bounds. This book is based on the premise that keeping things within bounds will be much, much easier if more of us understand the fundamental principles of the system organism–environment.

REFERENCES

1. Some philosophical thoughts on this subject from a recognized leader among biometeorologists may be found in F. Sargent, "The nature and nurture of biometeorology," *Amer. Inst. Biol. Sci. Bull.* **13**, 20

(1963), and identically in *Bull. Amer. Meteorol. Soc.* **44**, 483 (1963). Also, "A dangerous game: taming the weather," *Bull. Amer. Meterol. Soc.* **48** (7), 452 (1967).

2. Two clumsy but encouraging examples of the search for understanding total requirements come to mind: J. M. Caprio, *Agr. Meteorol.* **3**, 55, (1966) and W. P. Lowry, *Forest Sci.* **12**, 185 (1966).

Supplementary Reading List

These supplementary references are offered as having the following to commend them:

(a) They are of intermediate difficulty, and should be readily understood by anyone who has mastered the collateral materials in this text,
(b) they include further bibliographic items, many of them likely to be classics and most of them comparatively advanced and specialized,
(c) they include useful and informative data of many kinds, and
(d) they are reasonably accessible in the sense that most publishers and journals cited are domestic and well known.

Chapter 1

Amer. Meteorol. Soc. Study Group on Bioclimatology, F. Sargent II, Chairman, Biometeorology today and tomorrow, *Bull. Amer. Meteorol. Soc.* **48**, 378–393 (1967).

Chapters 2 through 6

H. R. Byers, "General Meteorology," McGraw-Hill, New York, 1959.

J. Edinger, "Watching for the Wind" (Anchor Sci. Study Ser. S 49), Doubleday, New York, 1967.

D. M. Gates, The energy environment in which we live, *Amer. Scientist* **51**, 327–438 (1963).

D. M. Gates, Energy, plants, and ecology, *Ecology* **46**, 1–13 (1965).

H. Lettau and B. Davidson, eds., "Exploring the Atmosphere's First Mile," 2 vols., Pergamon Press, New York, 1957.

R. E. Munn, Descriptive micrometeorology, *Advan. Geophys. Suppl.* **1** (1966).

C. H. B. Prestley, "Turbulent Transfer in the Lower Atmosphere," Univ. of Chicago Press, Chicago, Illinois, 1959.

W. E. Reifsnyder and H. W. Lull, Radiant energy in relation to forests, Tech. Bull. 1344, U.S. Dept. of Agriculture, Washington, D.C., 1965.

Chapters 7, 8, and 9

W. P. Lowry, Biometeological data collection, *Phytopathology* **53**, 1200–1202 (1963).

D. H. Miller, The heat and water budget of the earth's surface, *Advan. Geophys.* **11**, 175–302 (1965).

D. H. Miller, A survey course: the energy and mass budget at the surface of the earth, Comm. on Coll. Geography Publ. 7, Assoc. Amer. Geographers, Washington, D.C., 1968.

R. B. Platt, and J. Griffiths, "Environmental Measurement and Interpretation," Reinhold, New York, 1964.

W. D. Sellers, "Physical Climatology," Univ. of Chicago Press, Chicago, Illinois, 1965.

C. B. Tanner, Basic instrumentation and measurements for plant environment and micrometeorology, Soil Bull. No. 6, Dept. of Soil Sci., Univ. of Wisconsin, Madison, Wisconsin, 1963.

Chapters 10, 11, and 12

J. F. Bonner and A. W. Galston, "Plant Physiology," Freeman, San Francisco, California, 1952.

L. T. Evans, ed. "Environmental Control of Plant Growth," Academic Press, New York, 1963.

T. T. Kozlowski, "Water Deficits and Plant Growth," 2 vols., Academic Press, New York, 1968.

K. Raschke, Heat transfer between the plant and the environment, *Ann. Rev. Plant Physiol.* **11**, 111–126 (1960).

C. W. Rose, "Agricultural Physics," Pergamon Press, New York, 1966.

R. H. Shaw, ed., Ground Level Climatology, Publ. No. 86, Amer. Assoc. for the Advan. of Sci., Washington, D.C., 1967.

R. O. Slatyer, "Plant-Water Relationships," Academic Press, New York, 1967.

W. R. Van Wijk, ed., "Physics of Plant Environment," North-Holland Publ., Amsterdam, 1966.

Chapters 13 and 14

G. E. Folk, "Introduction to Environmental Physiology," Lea & Febiger, Philadelphia, Pennsylvania, 1966.

S. Licht, "Medical Climatology," Licht, New Haven, Connecticut, 1964.

W. P. Lowry, ed., Biometeorology, *Proc. 28th Ann. Biol. Colloq., Corvallis, Oregon, April 1967*, Oregon State Univ. Press, Corvallis, Oregon, 1968.

S. W. Tromp and W. H. Weihe, eds., "Biometeorology 2," 2 vols., Pergamon Press, New York, 1967.

Chapters 15 and 16

Air Conservation. Amer. Assoc. for the Advan. of Sci. Publ. No. 80, Amer. Assoc. for the Advan. of Sci., Washington, D.C., 1965.

L. J. Battan, "The Unclean Sky." (Anchor Sci. Study Ser. S. 46), Doubleday, New York, 1966.

W. P. Lowry, The climate of cities, *Sci. Amer.* **217**, No. 2, 15–23 (1967).

M. Neiburger, Weather modification and smog, *Science* **126**, 637–645 (1957).

D. H. Pack, Meteorology and air pollution, *Science* **146**, 1119–1128 (1964).

Chapter 17

W. D. Billings, "Plants and the Ecosystem," Wadsworth, Belmont, California, 1964.

H. E. Landsberg, Climate made to order, *Bull. At. Scientists* **17**, 370–374 (1961).

F. Sargent, II, A dangerous game: taming the weather, *Bull. Amer. Meteorol. Soc.* **48**, 452–458 (1967).

Solutions to Selected Problems

Chapter 3

3-1. 2.64:1.
3-5. (a) 355°K or 180°F. (b) 0.70. (c) 0.18.
3-6. 0.34.
3-7. (a) 0.86. (b) 1.28.
3-8. 64% increase.

Chapter 4

4-5. About 0.42.
4-6. (a) v_{sand} = about 0.86; ratio = 6:1. (b) $v_w = 0.14$; $v_{air} = 0.28$; $v_{gr} = 0.58$.
4-9. 240 cal.
4-10. (a) Pebble contains about 1525 cal. (b) Equivalent air volume about 21.7 liters. (c) 313°C.

Chapter 5

5-2. (a) 3.9 mb with condensation. (b) About 48%.

Chapter 6

6-1. 20°C.

Chapter 7

7-1. Hint: $-S_{uo} = t(S_{ui}) + r(S_{di})$ and similarly for S_{do} and S_{ho}.

Chapter 8

8-3. Sand: 0.52 cal cm^{-2} min^{-1}. Grass: 0.28 cal cm^{-2} min^{-1}. Conif.: 0.14 cal cm^{-2} min^{-1}.
8-4. About 80%.
8-5. About 0.04 cal cm^{-2} min^{-1}.
8-6. (a) 1.67 cal cm^{-2} min^{-1}. (b) 0.17×10^6 gm.
8-7. (a) 3.4 cal min^{-1}. (b) About 2°C.
8-8. About 2780 cm^2.

Chapter 10

10-5. About 61°F.

Chapter 11

11-2. (a) On Day 8. (b) In Year A.

Chapter 12

12-2. 6.7°C at $V = 50$; 2.7°C at 425.
12-3. About 65°C.
12-4. One burner per 180 meters^2 or about 42 feet apart.

Chapter 13

13-1. About 30°C.
13-2. During Day 6.

Chapter 14

14-2. 10.6 cal/cm/min/deg = 63.8 kcal/meter/hour/deg if $L_b = 0.5$ meter.
14-3. (a) An increase from about 7°C to about 9°C. (b) An increase from about 6°C to about 9°C.
14-4. (a) 25 HDD. (b) Day 7.

Chapter 15

15-2. Sand about 80°C rise; granite about 125°C.
15-4. (a) 0.35°C. (b) 1.0°C.

AUTHOR INDEX

SUBJECT INDEX

B

C

D